Parametric POMDPs

Alex Brooks

Parametric POMDPs

Planning in continuous spaces for mobile robot navigation

VDM Verlag Dr. Müller

Impressum/Imprint (nur für Deutschland/ only for Germany)
Bibliografische Information der Deutschen Nationalbibliothek: Die Deutsche Nationalbibliothek
verzeichnet diese Publikation in der Deutschen Nationalbibliografie; detaillierte bibliografische
Daten sind im Internet über http://dnb.d-nb.de abrufbar.
Alle in diesem Buch genannten Marken und Produktnamen unterliegen warenzeichen-, marken-
oder patentrechtlichem Schutz bzw. sind Warenzeichen oder eingetragene Warenzeichen der
jeweiligen Inhaber. Die Wiedergabe von Marken, Produktnamen, Gebrauchsnamen,
Handelsnamen, Warenbezeichnungen u.s.w. in diesem Werk berechtigt auch ohne besondere
Kennzeichnung nicht zu der Annahme, dass solche Namen im Sinne der Warenzeichen- und
Markenschutzgesetzgebung als frei zu betrachten wären und daher von jedermann benutzt
werden dürften.

Coverbild: www.purestockx.com

Verlag: VDM Verlag Dr. Müller Aktiengesellschaft & Co. KG
Dudweiler Landstr. 99, 66123 Saarbrücken, Deutschland
Telefon +49 681 9100-698, Telefax +49 681 9100-988, Email: info@vdm-verlag.de
Zugl.: Sydney, University of Sydney, 2007

Herstellung in Deutschland:
Schaltungsdienst Lange o.H.G., Berlin
Books on Demand GmbH, Norderstedt
Reha GmbH, Saarbrücken
Amazon Distribution GmbH, Leipzig
ISBN: 978-3-639-15627-0

Imprint (only for USA, GB)
Bibliographic information published by the Deutsche Nationalbibliothek: The Deutsche
Nationalbibliothek lists this publication in the Deutsche Nationalbibliografie; detailed
bibliographic data are available in the Internet at http://dnb.d-nb.de.
Any brand names and product names mentioned in this book are subject to trademark, brand or
patent protection and are trademarks or registered trademarks of their respective holders. The use
of brand names, product names, common names, trade names, product descriptions etc. even
without a particular marking in this works is in no way to be construed to mean that such names
may be regarded as unrestricted in respect of trademark and brand protection legislation and
could thus be used by anyone.

Cover image: www.purestockx.com

Publisher:
VDM Verlag Dr. Müller Aktiengesellschaft & Co. KG
Dudweiler Landstr. 99, 66123 Saarbrücken, Germany
Phone +49 681 9100-698, Fax +49 681 9100-988, Email: info@vdm-publishing.com
Sydney, University of Sydney, 2007

Printed in the U.S.A.
Printed in the U.K. by (see last page)
ISBN: 978-3-639-15627-0

To my parents, for making anything possible.

ii

Contents

iii

Nomenclature

Notation

$(\cdot)_k$	$(\cdot)$ at discrete time k
$(\cdot)^+$	$(\cdot)$ at the next time interval

General Symbols

k	an index into discrete time
$\mathbf{x}$	a state
X	the state space
$\mathbf{u}$	an action
U	the action space
f	the state transition function
$\mathbf{w}$	a disturbance affecting the state transition
T	a function describing the state transition probabilites
r	a state-based reward
R	a state-based reward function
γ	a discount factor
π	a policy
V	a value function
$\mathbf{z}$	an observation
Z	the observation space
h	the observation function
$\mathbf{v}$	a disturbance affecting the observation
O	the function describing the observation probabilites
c_0	the initial conditions available to a POMDP agent
$\mathbf{I}$	an information state
$\mathcal{I}$	an information space
$f_\mathbf{I}$	the information transition function
$r_\mathbf{I}$	an I-state-based reward
$R_\mathbf{I}$	the I-state-based reward function
$\pi_\mathbf{I}$	an information-based policy

$\mathcal{I}_{der}$	a derived information-space
$\mathbf{I}_{der}$	a derived information-state
κ_{der}	a mapping function, mapping to a derived information space
s	a discrete state
S	a set of discrete states
R_s	a reward function defined for discrete states
α	a value function hyperplane
Γ	a set of value function hyperplanes
$\Gamma^{\mathbf{u},\mathbf{z}^+}$	a set of intermediate alpha-vectors
λ	a weighting function
G	a set of states
$\mathbf{x}_{G,i}$	the i'th point in G
$\psi(\mathbf{x}_{G,i})$	the value of $\mathbf{x}_{G,i}$
Ψ	a set of state-value pairs
ϕ	a function approximator
B	a set of beliefs
$\mathbf{I}_{B,i}$	the i'th point in B
Δ	a set of posterior beliefs
$\mathbf{Q}$	a set of particles
q	a particle, *i.e.* a tuple of the form $<\mathbf{x}, w>$
w	a particle weight
$\hat{z}^+$	an expected observation
L	a likelihood matrix
W	a weight matrix
δ	the Dirac delta function
$\mathbf{z}_\alpha^+$	a component of the observation $\mathbf{z}^+$
Z_α	the space of possible values of $\mathbf{z}_\alpha^+$
$\mathbf{I}_\alpha^+$	an intermediate I-state produced after incorporating $\mathbf{z}_\alpha^+$
$\mathcal{I}_\alpha$	the space of possible values of $\mathbf{I}_\alpha^+$
$f_{\mathbf{I}_\alpha}$	the I-state transition function which incorporates $\mathbf{z}_\alpha^+$
B_α	a set of I-states in $\mathcal{I}_\alpha$
$\mathbf{I}_{B_\alpha,i}^+$	the i'th I-state in B_α, at the next time interval
T_α	a CPT governing transitions from B to B_α
$\mathbf{z}_C^+$	an observation from a robot's collision sensor
$\mathbf{z}_R^+$	an observation from a robot's range sensors
D	a distance measure
P	a database of points
V	the domain of P

q	a query into the database
r	the radius of a search query
$\mathbf{v}$	a vector of parameters
τ	the distance to the closest point found so far during a search
η	a kernel function
ζ	a bandwidth
F	a set of features
$\mathbf{f}_i$	the i'th feature in F
Z_f	a set of polar feature observations
$\mathbf{z}_{f,i}$	the i'th feature observation in Z_f, of the form $< z_{r,i}, z_{b,i} >$
$z_{r,i}$	the range of $\mathbf{z}_{f,i}$
$z_{b,i}$	the range of $\mathbf{z}_{f,i}$
$\hat{Z}_f$	the expected set of feature observations

Abbreviations

MDP	Markov Decision Process
POMDP	Partially Observable Markov Decision Process
FVI	Fitted Value Iteration
CPD	Conditional Probability Distribution
CPT	Conditional Probability Table
SLAM	Simultaneous Localisation and Mapping
PWLC	Piecewise-Linear and Convex

Chapter 1

Introduction

This book is concerned with the problem of planning and acting in uncertain, partially observable, continuous domains. In particular, it focusses on the task of planning and acting for mobile robot navigation when a map of the environment is available. Robot navigation problems are particularly challenging for planners because they are inherently continuous, uncertain, and non-linear. However, the ability to make good plans despite these conditions is fundamental to an autonomous mobile robot's ability to navigate reliably in real-world environments.

Classical Artificial Intelligence (AI) planning assumes that environments are fully-observable, deterministic, finite, static, and discrete [95]. The first major planning system for such environments was STRIPS [39], which represented the state of the environment with a set of symbols. A set of actions were posited, each of which had a set of pre-conditions and a set of deterministic effects on the symbolic state of the world. Given a start state, the definitions of actions, and a goal state, a STRIPS-style planner could autonomously map out a fixed sequence of actions which would lead to that goal.

Unfortunately, few of the assumptions of classical AI planning hold for realistic mobile robot applications. Fixed sequences of actions are inappropriate because actions' outcomes are unpredictable. Real robots therefore have difficulty executing STRIPS-style plans [40]. Instead, feedback is required: an agent must observe the world and react accordingly. A number of extensions allow classical AI planning systems to incorporate feedback, for example by making plans conditional on the state of the world [95][12].

Another important omission of classical AI planning is that, rather than reaching one or more goal states, agents in realistic scenarios are usually required to satisfy various (possibly competing) objectives simultaneously. For example, robots should act so as to minimise the risk of encountering hazards which might cause them harm. One way of specifying objectives is through the use of a reward[1] function [105][59], which specifies the desirability of possible states

---

[1] While reward is usually used in the computer science literature, cost (negative reward) is usually used in

of the world, and perhaps the desirability of particular actions in particular states. A more sophisticated model, which accounts for unpredictable actions and general reward functions, is a Markov Decision Process.

1.1   Markov Decision Processes

A Markov Decision Process, or MDP, provides a general mathematical model for the interaction between an agent and the world. Many classical AI planning algorithms can be formulated as special cases of MDPs [16]. An MDP assumes that the state of the world at any time can be described by a set of continuous or discrete variables. This state evolves in small discrete time-steps, affected by the agent's actions. The agent chooses these actions based on its direct and infallible knowledge of the state.

To account for un-predictability in the world, the MDP model requires that the effects of agents' actions can be described by stationary probability distributions. That is, from any given state and for any given action, an MDP specifies a probability distribution over subsequent states.

Matters are simplified considerably by the Markov assumption [111]. This asserts that the current state is a sufficient statistic for the past. In other words, if the agent knows the current state of the world, the details of how the world came to be in that state convey no extra information about what will happen in the future. This is usually a fairly accurate assumption for the real world, given a sufficiently descriptive state vector.

By framing a problem as an MDP, one gains access to a powerful arsenal of solution algorithms [16][105][12]. The output of many solution algorithms is a value function, which specifies a value for every possible state. Loosely speaking, this value is the sum of rewards which can be obtained in the future by acting optimally (with an infinite-horizon lookahead) from that state. Armed with the value function, an agent need not plan ahead, since plans are implicitly encoded in the value function. An agent can act optimally simply by greedily choosing actions which will immediately lead to high-value states.

MDP solution algorithms have solved many challenging problems, particularly for board games such as backgammon [106]. Translating this success to real-world problems can be problematic however, because the MDP formulation assumes that the agent has perfect knowledge of the state. A more realistic formulation assumes only partial observability.

---

the economics and operations research literature [10]

1.2 Partial Observability

While an MDP models uncertainty in an agent's actions, it assumes that the agent is completely aware of the state of the world. This assumption is often invalid in real scenarios, particularly for the kinds of problems considered in this book. A more realistic model is a Partially Observable Markov Decision Process, or POMDP. A POMDP extends an MDP by assuming that, rather than sensing the state of the world directly, an agent can make observations which give it imperfect information about the state. The POMDP formulation assumes that the likelihood of observations given the hidden state can be described by stationary probability distributions.

While an MDP agent has the luxury of making decisions based on the state, a POMDP agent must make decisions based solely on the history of actions and observations. This history represents everything the agent knows about the world, and is often referred to as the information-state. This dependence on history complicates matters. For MDPs, the Markov property implies that an agent can safely ignore history. The POMDP formulation also assumes that the state obeys the Markov property, however this state is no longer directly observable. The entire history of actions and observations is therefore relevant as it potentially confers information about the hidden state.

The dominant approach to avoiding this history is to use all the available information to maintain a probability distribution, or belief, over possible states. The belief is a sufficient statistic for history. That is, if an agent knows the current probability distribution, the observations and actions which led to that distribution are irrelevant for predicting the future. It will be shown in Chapter 2 that a POMDP can be seen as a special kind of MDP. The unobservable state can be replaced by the observable information-state, which can be summarised by a belief-state. Standard MDP solution algorithms can then be applied to the resultant MDP.

POMDPs are excellent models for many mobile robot navigation problems in which the state is the pose (position and heading) of the robot. A typical scenario involves the use of sensors such as cameras, laser range-finders, and wheel encoders to gather information. The information is imperfect because sensors are noisy, cannot look everywhere at once, and usually cannot sense the state of interest (*i.e.* the pose) directly. In this context, a POMDP solution represents a plan which allows the robot to gather the information it requires while simultaneously bringing it to its goal.

1.3 Solving POMDPs

Realistic mobile navigation problems are difficult to solve using POMDPs because their state, action, and observation spaces are large and continuous. It will be shown in Chapter 2 that

Figure 1.1: POMDP solution methods span a continuum. On one extreme, the combination of a fine discretisation with exact value iteration will produce excellent plans but will be incapable of scaling to realistic problems. On the other extreme, heuristics require little computation but fail to take a principled approach to uncertainty. The most useful planner lies somewhere in the middle. The figure is adapted from [94].

a continuous state-space implies an infinite-dimensional continuous value function. Since this cannot be represented except in very special cases, approximations are clearly required for the general case. Solving POMDPs for robot navigation problems is therefore a game of approximations. One must strike the right balance between approximations which over-simplify the problem to the point where the robot is incapable of planning effectively, and approximations which do not simplify the problem enough, leaving it computationally intractable. This trade-off is illustrated in Figure 1.1.

The simplest approach is to choose actions based on the assumption that the most likely pose is true, ignoring uncertainty. As an improvement, a number of heuristics for dealing with uncertainty have been devised. The details are discussed in Section 2.7.1, however suffice it to say that these approaches can only reason about the evolution of probability distributions over a very short planning horizon. In terms of Figure 1.1 they err on the side of over-simplification of the problem, producing plans which are not robust in the presence of uncertainty.

Another common approximation is to discretise the state, action and observation spaces. Having done so, a number of exact solution algorithms exist such that no further approximation is required [23]. Unfortunately, these exact algorithms are considered to be incapable of scaling to real-world problems in general, and will certainly not scale to the kinds of problems considered in this book. They lie on the opposite extreme of the spectrum depicted in Figure 1.1, representing an under-simplification of the problem.

After applying discretisation, a number of further approximations stem from the important insight that not all probability distributions are equally relevant. Figure 1.2 shows a hypothetical example of two probability distributions for a robot navigation problem. The belief shown in Figure 1.2(a) is certainly relevant, in that it is typical of the kinds of distributions that are expected to be encountered in practice. If the robot's poor planning has not considered this belief, that poor planning is likely to be exposed. In contrast, Figure 1.2(b) shows an irrelevant belief. It is impossible, or at least highly unlikely, for this belief to occur. Hence, a robot which has specifically planned for this belief is unlikely to perform any differently from a robot which

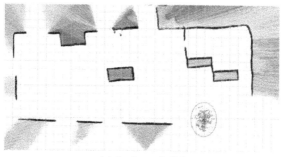

(a) A relevant belief

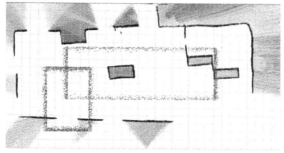

(b) An irrelevant belief

Figure 1.2: A relevant and an irrelevant belief. Beliefs are represented as particle sets, shown in blue. Particle density is proportional to probability density. The belief shown in (a) is relevant because similar beliefs are likely to occur during plan execution. Since it is impossible or at least highly unlikely that the belief in (b) will occur, a planner will only waste time by considering it. The beliefs are displayed on top of an occupancy grid [37]. Black denotes an occupied cell, white denotes an empty cell, and grey denotes unknown occupancy. Adapted from [94].

has ignored it.

This insight has been used in two ways:

1. to focus computation on a set of likely beliefs within a class, and

2. to restrict the class of beliefs which can be considered.

The first approach includes point-based methods which generate belief sets by model simulation [86][103][101]. Using some policy (*e.g.* random actions [103]), these methods simulate the repeated interaction of an agent with the POMDP model in order to generate a representative set of likely beliefs. Computation can then be focussed on these beliefs during planning. The hope is that the plan will generalise from this representative set of beliefs to all beliefs which are likely to be encountered in practice.

The second approach is to restrict the class of beliefs which can be considered. This includes approaches such as belief compression [94], which is based on dimensionality reduction. The space of all possible probability distributions over a discrete set of states is high-dimensional and continuous. Rather than allowing arbitrary beliefs, belief compression restricts itself to those which lie on a low-dimensional manifold embedded in that high-dimensional space. By choosing the manifold carefully, the set of beliefs which are likely to occur in practice will hopefully lie on or near that manifold.

The approach advocated in this book can be seen as a case of restricting the class of representable beliefs. The important difference from previous work is that we do not begin with discretisation. Instead, we assume that beliefs can be approximated by continuous functions described by finite sets of parameters. The space of functions prescribes the class of beliefs to which the planner is restricted. If this space is chosen appropriately, it will hopefully be a good approximation to the kinds of beliefs which will occur in practice.

We will show that this difference has important ramifications, allowing us to scale to real-world problems. Most importantly, the use of parameterised continuous functions provides a compact representation of beliefs which does not rely on an underlying discretisation. For large problems, an underlying discrete representation is problematic. One must choose between a fine discretisation, which introduces scalability problems, and a coarse discretisation, with which one is unable to represent smooth gradients and small shifts in distributions.

In contrast, the use of continuous parameterised functions can handle smooth gradients and small shifts, but introduces a choice of function complexity. The use of overly complex functions may cause a problem to be intractable, whereas simple functions may constrain the shapes of beliefs too tightly, resulting in poor plans. We will show that for robot navigation problems, Gaussians represent a class of functions which are sufficiently simple to allow us to scale to large problems. At the same time they are sufficiently expressive to closely approximate the kinds of probability distributions which are usually, but not always, encountered during robot navigation. Section 3.1 will discuss the validity of this approximation in detail.

1.4   Application Domain

The particular application which this book works towards is the reliable operation of the autonomous mobile robot shown in Figure 1.3 in a large semi-structured outdoor environment. The robot senses the environment with a forward-looking laser range-finder, and senses its

Figure 1.3: A robot and its environment. The robot is dynamically stabilised. A forward-looking scanning laser range-finder is mounted centrally, just above the wheels. A second robot is in the background.

own motion with wheel encoders. We assume that the robot is given an *a priori* map of the environment.

The scenario considered here is particularly challenging for a number of reasons. Debris and holes in the asphalt make odometry particularly poor. Since the robots are statically unstable, they must pitch back and forth in order to balance, especially when accelerating or decelerating. This makes sensing with a fixed laser complicated. On the edge of the robots' working area lies natural terrain for which an accurate geometric model would be extremely complicated. In addition, the environment contains a number of hazards which could cause the robots to fall. These hazards lie below the plane of the laser, and are therefore essentially invisible to the robots.

Indeed, this is the most challenging robot navigation problem to which POMDP solution methods have been applied, by a significant margin. We would argue that Roy's work represents the most challenging problem previously attempted [94]. This involved a simulated environment of a similar size. The problem was simplified by using an omni-directional sensor and ignoring the robot's heading. The addition of heading is more realistic but much more challenging, adding an extra dimension to the problem. By excluding heading from the POMDP model, the robot's actions must be specified in absolute terms, which is unrealistic unless the robot is somehow aware of its absolute heading during plan execution. Furthermore, the robot cannot be aware of the fact that sharp turns now increase uncertainty in the future. The simplification of using an omni-directional sensor means that the robot needn't consider the fact that its ability to gather information depends on the direction in which it is travelling. POMDP solution methods which rely on this simplification are unable to generalise to the arguably more common case of sensors which are not omni-directional. In contrast, our application includes heading and uses a forward-pointing sensor, implying that our robot must be able to account for all these details.

A solution to the decision-making problem relies on a robust solution to the localisation problem. That is, the robot must be able to use its uncertain sensor readings to work out where in the map it might be. While this is challenging, it is by no means unsolvable. Excellent progress has been made in this area over the last couple of decades by casting the problem as one of Bayesian estimation, and by applying approximations which make the problem computationally tractable in real-time [111]. Essentially, the robot maintains a probability distribution (or belief) over possible poses, and uses the actions and observations at each time interval to update this probability distribution.

Unfortunately, while the agent has access to these powerful methods of maintaining probability distributions, practical systems do not generally use the entirety of those distributions for decision-making. The standard approach is to assume that the most likely pose is in fact the true pose. To get from a start location to a destination, one can then apply any one of a number of deterministic path-planning algorithms which assume complete observability and deterministic actions [63][64]. This blind faith that the most likely pose is true can lead the robot to be overly confident, with potentially catastrophic results. POMDPs provide a framework for overcoming this problem by using the entire probability distribution, however the computational complexity of solving POMDPs has prohibited their widespread adoption. The following section outlines the contributions this book makes in order to address this issue.

1.5   Contributions

The contributions of this book are as follows:

- The presentation of a unifying view of several POMDP solution methods from the literature as specific instances of a more general solution method, namely the application of fitted value iteration in a particular information-space.

- The development and analysis of a novel planning algorithm, entitled PPOMDP, representing a specific instance of the general methodology defined above. PPOMDP entails the application of fitted value iteration in the space of continuous parameterised functions.

- The formulation of an approach to estimating distributions over posterior parameterised beliefs using methods from the particle filtering literature. This novel approach is shown to scale independently from the size of the state-space, and hence is applicable to large, realistic planning problems.

- The formulation of a simplification of the planning problem using a factoring based on conditional independence assumptions. With certain approximations, this novel approach allows algorithms based on fitted value iteration to be broken into smaller components, reducing the total computational complexity.

- The presentation of a method for efficient function approximation for arbitrary sets of parameterised beliefs, using data structures from the similarity search literature.

- The efficient integration of local online forward planning into the PPOMDP framework, assisting offline global planning.

- The experimental evaluation of the PPOMDP approach in its various forms, and an experimental comparison against a state-of-the-art POMDP solution algorithm, in several simulated environments.

- The real-time implementation and experimental validation of the PPOMDP algorithm on a real robot navigating in a challenging environment. To the author's knowledge, this work represents the most challenging robot navigation problem to which POMDP solution methods have successfully been applied to date.

1.6 Book Structure

Chapter 2 introduces basic MDP and POMDP terminology and concepts more formally, then reviews numerous solution algorithms which have been proposed in the literature. Chapter 3 introduces the basic concepts and solution algorithm behind the approach in this book, namely planning in the space of parameterised continuous functions. It argues for the use of Gaussian functions to approximate beliefs encountered in robot navigation problems. In order to compare approaches, Chapter 3 introduces BlockWorld: a simple continuous navigation problem. We describe the details necessary to implement our algorithm for BlockWorld, then compare its performance against an MDP-based heuristic and a state-of-the-art point-based algorithm [103]. The results and algorithm presented in Chapter 3 are very similar to previously published work [19].

Chapter 3 serves as a foundation for the remaining chapters of this book. The algorithm presented in Chapter 3 has a number of deficiencies, in terms of both performance and scalability.

Subsequent chapters maintain the same basic approach, but present improvements to both the quality of plans and the scalability of planning, to the point where the algorithm can operate competently in real environments. Each of Chapters 4 through 7 build on the algorithm as presented in the previous chapter by improving on a specific aspect, and present results on the BlockWorld problem to quantify that improvement.

Chapter 4 highlights some of the deficiencies of the Algorithm from Chapter 3. It suggests an improved and more general algorithm for projecting beliefs forward in time, and describes details of how it can be implemented efficiently. This improvement allows the algorithm to produce significantly better plans in approximately the same amount of time.

In Chapter 5, it will be shown how planning speed can be improved dramatically by pre-calculating the effects of observations. Essentially, the problem can be broken down into smaller components by a factoring based on conditional independence assumptions. This improvement produces similar results to the algorithm presented in the previous chapter, but in a fraction of the time.

Having already restricted the space of representable beliefs, Chapter 6 focusses computation on the important areas of that space. The algorithm requires a set of sample beliefs to plan over. Until this point, the algorithm has required that these lie on a regular grid over belief-space. In order to relax this requirement, a method is needed to efficiently retrieve the set of beliefs in the vicinity of a query belief. Chapter 6 reviews data structures from the similarity search literature and applies them to achieve this aim, resulting in a significant increase in scalability.

Chapter 7 introduces the final improvement. It shows how real-time online planning can be integrated with the offline planning algorithm described in previous chapters. Essentially, this allows an online agent to locally "fill in the gaps" of an occasionally coarse pre-computed global plan.

Chapter 8 moves beyond BlockWorld and applies the algorithm, with all its improvements, to a real problem. It describes the environment outlined in Section 1.4 in more detail, and explains how the algorithm presented in previous chapters can be applied to it. It presents results first on a toy simulated world with realistic dynamics, then on a simulated version of the real environment, and finally on a real robot operating in the real environment. Chapter 9 concludes and discusses future work.

Chapter 2

Sequential Decision Making

Partially observable Markov decision processes (POMDPs) provide a general mathematical framework for modelling problems involving sequential decision making in partially observable domains. Exact solutions to POMDP problems allow an agent to act optimally in the presence of uncertainty. Many real-world problems, including robot navigation, are well modelled by POMDPs. This chapter begins by establishing some basic terminology and concepts for both fully observable and partially observable Markov decision processes. It shows how the partially observable state can be replaced with the fully observable history of all information available to the agent. Various approaches to representing that history compactly are discussed, and numerous algorithms for solving the resultant decision-making problems are reviewed.

2.1 Markov Decision Processes

A Markov Decision Process (MDP) involves a decision-making agent interacting with a fully observable stochastic environment, as shown in Figure 2.1. The environment includes the

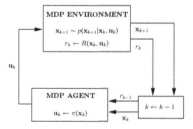

Figure 2.1: The MDP model. At each iteration the agent produces a new action $\mathbf{u}_k$ based on the state. The world samples a new state $\mathbf{x}_{k+1}$ based on the agent's action. The "$k \leftarrow k - 1$" box simply alters time subscripts in preparation for the next iteration.

entirety of the decision maker's world. For robot navigation problems, this includes the robot's pose. Throughout this document it is assumed that time is discretised into a set of intervals indexed by k. The state vector $\mathbf{x}_k \in X$ is used to describe the state of the environment at time k. The Markov property asserts that the state is a sufficient statistic for history, meaning that the past conveys no extra information about the future if the present is known [111]. At each time interval the agent chooses an action $\mathbf{u}_k \in U$ which causes the state to transition stochastically from $\mathbf{x}_k$ to $\mathbf{x}_{k+1}$, and results in the agent receiving reward r_k. This book assumes that the agent has complete knowledge of the probabilistic model used for state transitions.

Formally, an MDP is defined by the tuple

$$< X, U, T, R, \mathbf{x}_0, \gamma > \tag{2.1}$$

where

1. X is the state-space;

2. U is the space of actions;

3. $T(\mathbf{x}_k, \mathbf{u}_k, \mathbf{x}_{k+1}) = p(\mathbf{x}_{k+1}|\mathbf{x}_k, \mathbf{u}_k)$ defines transition probabilities between states;

4. $R(\mathbf{x}_k, \mathbf{u}_k)$ defines a reward function;

5. $\mathbf{x}_0$ is the initial state of the environment; and

6. γ is a discount factor.

Over an episode, the agent executes a policy π, which specifies an action for every state:

$$\pi : X \rightarrow U \tag{2.2}$$

This policy can be seen as a conditional plan. After the first action is taken, subsequent actions depend on the outcomes of stochastic state transitions.

Let $V_\pi(\mathbf{x}_k)$ denote the value of executing policy π starting from state $\mathbf{x}_k$. $V_\pi(\mathbf{x}_k)$ is referred to as a value function, and is equal to the discounted sum of expected future rewards:

$$V_\pi(\mathbf{x}_k) = \sum_{j=0}^{K} \gamma^j \underset{\mathbf{x}_{k+j}}{E} \left[R\big(\mathbf{x}_{k+j}, \pi(\mathbf{x}_{k+j})\big) \right] \tag{2.3}$$

where γ is a discount factor ≤ 1 and K is the time remaining in the episode. This document considers only the discounted infinite horizon case, where $K = \infty$ and $\gamma < 1$.[1] Equation 2.3

---

[1]The case of $K = \infty$ and $\gamma = 1$ is the subject of ongoing research [105].

calculates the value of a policy based on all future rewards. These future rewards depend on future states. While future states are unknown, the probabilistic model of the environment can be used to take an expectation over future states. The discount factor γ is used to weight rewards in the near future more heavily than rewards in the distant future. This simplifies matters by inducing finite values.

The Bellman equation is a recursive version of Equation 2.3, defining the value function at time k recursively in terms of the value function at time $k + 1$ [10][105]:

$$V_\pi(\mathbf{x}_k) = R(\mathbf{x}_k, \mathbf{u}_k) + \gamma \underset{\mathbf{x}_{k+1}}{E} \left[V_\pi(\mathbf{x}_{k+1}) | \mathbf{u}_k \right] \tag{2.4}$$

$$= R(\mathbf{x}_k, \mathbf{u}_k) + \gamma \int_{\mathbf{x}_{k+1}} V_\pi(\mathbf{x}_{k+1}) p(\mathbf{x}_{k+1} | \mathbf{x}_k, \mathbf{u}_k) d\mathbf{x}_{k+1} \tag{2.5}$$

That is, the value of executing a policy from a state is the immediate reward plus a discounted version of the future reward. The future reward is the (possibly infinite) sum of the values of all possible next-states, weighted by their probability.

The aim of the MDP agent is to find the optimal policy

$$\pi^*(\mathbf{x}_k) = \arg\max_\pi V_\pi(\mathbf{x}_k) \tag{2.6}$$

Combining this with Equation 2.4 gives

$$V_{\pi^*}(\mathbf{x}_k) = \max_{\mathbf{u}_k} \left[R(\mathbf{x}_k, \mathbf{u}_k) + \gamma \underset{\mathbf{x}_{k+1}}{E} \left[V_{\pi^*}(\mathbf{x}_{k+1}) | \mathbf{u}_k \right] \right] \tag{2.7}$$

Many decision-making algorithms find policies through exact or approximate solutions to this equation [105]. The solution algorithm on which this book focusses is value iteration.

2.1.1 Solving MDPs Using Value Iteration

While solution methods for continuous MDPs remain an open problem [64], MDPs with discrete state, action and observation spaces are in principle relatively straightforward to solve using value iteration. Equation 2.7 can be re-written, replacing the expectation with a summation:

$$V_{\pi^*}(\mathbf{x}_k) = \max_{\mathbf{u}_k} \left[R(\mathbf{x}_k, \mathbf{u}_k) + \gamma \sum_{\mathbf{x}_{k+1} \in X} V_{\pi^*}(\mathbf{x}_{k+1}) T(\mathbf{x}_k, \mathbf{u}_k, \mathbf{x}_{k+1}) \right] \tag{2.8}$$

For relatively small discrete MDPs, a simple approach to evaluating Equation 2.8 is to represent R, T and V explicitly, for every state and action, using a set of tables. With this representation, the discrete MDP can be solved as shown in Algorithm 1. For larger MDPs which cannot be represented explicitly in tables, a number of approximations to Algorithm 1 exist [105].

Algorithm 1 A solution algorithm for relatively small discrete MDPs represented explicitly by tables of rewards, transition probabilities and values. For each state, the algorithm outputs both the value and the best action. A typical convergence criterion is when the maximum change in the value function drops below a threshold.

---

1    $V_k(\mathbf{x}_k) \leftarrow 0, \forall \mathbf{x}_k \in X$
2   *while not* **converged**
3       $V_{k+1} \leftarrow V_k$
4   *forall* $\mathbf{x}_k \in X$
5   *forall* $\mathbf{u}_k \in U$
6           $V_{\mathbf{u}}(\mathbf{u}_k) \leftarrow R(\mathbf{x}_k, \mathbf{u}_k) + \gamma \sum_{\mathbf{x}_{k+1} \in X} T(\mathbf{x}_k, \mathbf{u}_k, \mathbf{x}_{k+1}) V_{k+1}(\mathbf{x}_{k+1})$
7   *end forall*
8         $\mathbf{u}_{best} \leftarrow \arg\max_{\mathbf{u}_k} \left[V_{\mathbf{u}}(\mathbf{u}_k) \right]$
9       $V_k(\mathbf{x}_k) \leftarrow V_{\mathbf{u}}(\mathbf{u}_{best})$
10   *end forall*
11 *end while*

Algorithm 1 can be seen as iteratively stepping backward in time. It assumes a value function, defined for all states. Each iteration then steps back one step in time, and constructs a one-step policy to reach that (now future) value function. Based on this policy and the future value function, the agent can build a value function for the present. Convergence occurs because the discount factor γ causes changes in the value function to diminish geometrically. The MDP agent can actually execute the policy without reference to the value function, simply by remembering the maximising actions from Step 8.

Algorithm 1 is simple but fundamental. The majority of this book examines efficient methods for manipulating POMDPs such that they can be solved using Algorithm 1.

2.2   From MDPs to POMDPs

The inclusion of partial observability implies that the state of the environment is not directly available to the agent. Instead, at each interval k the agent receives an observation $\mathbf{z}_k \in Z$ which confers incomplete information regarding the state.

More formally, a POMDP is defined by the tuple

$$< X, U, Z, T, O, R, c_0, \gamma > \tag{2.9}$$

where

1. X is the state-space;

2. U is the space of actions;

3. Z is the space of observations;

4. $T(\mathbf{x}_k, \mathbf{u}_k, \mathbf{x}_{k+1}) = p(\mathbf{x}_{k+1}|\mathbf{x}_k, \mathbf{u}_k)$ defines transition probabilities between states;

5. $O(\mathbf{x}_{k+1}, \mathbf{u}_k, \mathbf{z}_{k+1}) = p(\mathbf{z}_{k+1}|\mathbf{x}_{k+1}, \mathbf{u}_k)$ defines observation probabilities;

6. $R(\mathbf{x}_k, \mathbf{u}_k)$ defines a reward function;

7. c_0 is the initial information available to the agent; and

8. γ is a discount factor.

It is assumed that the initial information c_0 takes the form of a (possibly uniform) probability distribution over state-space.

While many approaches exist for solving POMDPs, the most prevalent (and the one on which this book focusses) is value iteration. Value iteration can be applied by viewing the POMDP as an information-state MDP.

2.2.1 POMDPs as Information-State MDPs

An MDP agent's environment can be considered to consist of two components: an observable deterministic world model which receives an unobservable non-deterministic disturbance [10]. The non-deterministic disturbance, denoted $\mathbf{w}_{k+1}$, is selected from the probability distribution $p(\mathbf{w}_{k+1}|\mathbf{x}_k, \mathbf{u}_k)$, which the agent is assumed to know. Given this input, the world evolves according to a deterministic state transition function f, such that $\mathbf{x}_{k+1} = f(\mathbf{x}_k, \mathbf{u}_k, \mathbf{w}_{k+1})$. This interpretation is depicted in Figure 2.2(a).

As a concrete example, additive white Gaussian process noise is often assumed (*e.g.* in the Kalman filtering literature). In this case, $p(\mathbf{w}_{k+1}|\mathbf{x}_k, \mathbf{u}_k)$ is a Gaussian distribution. The transition function $f(\mathbf{x}_k, \mathbf{u}_k, \mathbf{w}_{k+1})$ applies the action $\mathbf{u}_k$ deterministically, then adds the Gaussian perturbation $\mathbf{w}_{k+1}$ to the resultant state.

Together, $p(\mathbf{w}_{k+1}|\mathbf{x}_k, \mathbf{u}_k)$ and f determine the state transition probabilities $T(\mathbf{x}_k, \mathbf{u}_k, \mathbf{x}_{k+1}) = p(\mathbf{x}_{k+1}|\mathbf{x}_k, \mathbf{u}_k)$:

$$p(\mathbf{x}_{k+1}|\mathbf{x}_k, \mathbf{u}_k) = \int_{\mathbf{w}_{k+1}} f(\mathbf{x}_k, \mathbf{u}_k, \mathbf{w}_{k+1}) p(\mathbf{w}_{k+1}|\mathbf{x}_k, \mathbf{u}_k) d\mathbf{w}_{k+1} \qquad (2.10)$$

Similarly, for the POMDP case, the observation function $O(\mathbf{x}_{k+1}, \mathbf{u}_k, \mathbf{z}_{k+1}) = p(\mathbf{z}_{k+1}|\mathbf{x}_{k+1}, \mathbf{u}_k)$ can be seen as consisting of two components: a non-deterministic observation disturbance $\mathbf{v}_{k+1}$ drawn from $p(\mathbf{v}_{k+1}|\mathbf{x}_{k+1}, \mathbf{u}_k)$ and a deterministic observation function $h(\mathbf{x}_{k+1}, \mathbf{u}_k, \mathbf{v}_{k+1})$. Note

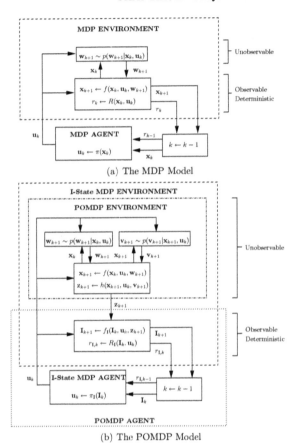

Figure 2.2: The parallel between the (a) MDP and (b) POMDP models. The I-state MDP agent plays the same role as the MDP agent, but $\mathbf{x}$ and r are replaced by $\mathbf{I}$ and $r_{\mathbf{I}}$. Both the MDP and I-state MDP agents' environments consist of an observable deterministic component which receives a stochastic input from an unobservable component. Despite this source of randomness, planning is possible in MDPs and POMDPs because agents can model the probabilities of future disturbances and future observations respectively.

that, while we allow for the more general case, it is usually assumed that both $\mathbf{w}_{k+1}$ and $\mathbf{v}_{k+1}$ are independent of the action.

To continue with the example, models in the Kalman filtering literature use additive white Gaussian observation noise. Under this assumption, $p(\mathbf{v}_{k+1}|\mathbf{x}_{k+1}, \mathbf{u}_k)$ is a Gaussian and $h(\mathbf{x}_{k+1}, \mathbf{u}_k, \mathbf{v}_{k+1})$ simply adds $\mathbf{v}_{k+1}$ to the observation.

Without direct access to the state, the POMDP agent must make decisions based on the available information. The set of all available information at time k consists of

1. *a priori* knowledge of the initial state of the world, denoted $c_0 \in C_0$;

2. all observations up to and including time k; and

3. all actions up to and including time $k - 1$.

A representation for this set of information is termed an information-state, or I-state for brevity, denoted $\mathbf{I}_k$. $\mathbf{I}_k$ can be viewed as a point in the space of all possible I-states, termed the information space (or I-space) $\mathcal{I}$ [112][64]. Particular representations of $\mathbf{I}_k$ are discussed in detail later in this chapter, but by definition it is completely observable.

In order to base its decisions on I-states, the POMDP agent needs to know how actions and observations will modify those I-states. This is specified by an I-state transition function, denoted $f_\mathbf{I}$, such that

$$\mathbf{I}_{k+1} = f_\mathbf{I}(\mathbf{I}_k, \mathbf{u}_k, \mathbf{z}_{k+1}) \tag{2.11}$$

In addition, in the absence of direct access to the reward (which would give the agent clues as to the true state of the world), the POMDP agent must use an estimate of its reward, denoted $r_\mathbf{I}$, obtained through the I-state-based reward function $R_\mathbf{I}$:

$$r_\mathbf{I} = R_\mathbf{I}(\mathbf{I}_k, \mathbf{u}_k) \tag{2.12}$$

Given these functions, the POMDP can be viewed as an I-state MDP by replacing the unobservable state of the environment $\mathbf{x}$ with the fully-observable I-state $\mathbf{I}$, and replacing the state-based reward r with the I-state-based reward $r_\mathbf{I}$, as illustrated in Figure 2.2(b). Furthermore, there is a parallel with the view of an MDP agent's environment as an observable component receiving a stochastic disturbance from an unobservable component. In the POMDP case, from the point of view of the I-state MDP agent, the observable deterministic component consists of the I-state transition and reward functions. The unobservable component is the true environment, and the stochastic disturbance is the observation. Just as the MDP agent has knowledge of the distribution $p(\mathbf{w}_{k+1}|\mathbf{x}_k, \mathbf{u}_k)$, the POMDP agent can construct the distribution $p(\mathbf{z}_{k+1}|\mathbf{I}_k, \mathbf{u}_k)$ based on its knowledge of the world model and the two true sources of non-determinism, $p(\mathbf{w}_{k+1}|\mathbf{x}_k, \mathbf{u}_k)$ and $p(\mathbf{v}_{k+1}|\mathbf{x}_{k+1}, \mathbf{u}_k)$. This does not imply that $\mathbf{z}_{k+1}$ actually depends on $\mathbf{I}_k$, but that $\mathbf{I}_k$ tells the agent something about the distribution over $\mathbf{z}_{k+1}$.

The POMDP agent's task is to optimise an information-based policy $\pi_\mathbf{I}$ which specifies an action for every point (I-state) in the I-space:

$$\pi_\mathbf{I} : \mathcal{I} \rightarrow U$$

	MDP	I-State MDP
Observable "State"	$\mathbf{x}_k$	$\mathbf{I}_k$
Reward	r_k	$r_{\mathbf{I},k}$
Policy	π	$\pi_{\mathbf{I}}$
Disturbance	$\mathbf{w}_{k+1}$	$\mathbf{z}_{k+1}$
Reward Model	$R(\mathbf{x}_k, \mathbf{u}_k)$	$R_{\mathbf{I}}(\mathbf{I}_k, \mathbf{u}_k)$
Transition Function	$f(\mathbf{x}_k, \mathbf{u}, \mathbf{w}_{k+1})$	$f_{\mathbf{I}}(\mathbf{I}_k, \mathbf{u}_k, \mathbf{z}_{k+1})$
Disturbance Model	$p(\mathbf{w}_{k+1}|\mathbf{x}_k, \mathbf{u}_k)$	$p(\mathbf{z}_{k+1}|\mathbf{I}_k, \mathbf{u}_k)$

Table 2.1: The equivalence between various quantities in either an MDP or a POMDP viewed as an I-state MDP. For an I-state MDP, the quantities below the line can be derived from the definition of the I-state and the mechanics of the underlying MDP.

Adapting Equations 2.3 and 2.6 to the new information-state MDP results in the definitions of the value of an I-state:

$$V_{\pi_{\mathbf{I}}}(\mathbf{I}_k) = \sum_{j=0}^{K} \gamma^j \underset{\mathbf{I}_{k+j}}{E} \left[R_{\mathbf{I}}\big(\mathbf{I}_{k+j}, \pi_{\mathbf{I}}(\mathbf{I}_{k+j})\big) \right] \tag{2.13}$$

and the optimal information-based policy:

$$\pi_{\mathbf{I}}^*(\mathbf{I}_k) = \arg\max_{\pi_{\mathbf{I}}} V_{\pi_{\mathbf{I}}}(\mathbf{I}_k) \tag{2.14}$$

The information-based analogue of Equation 2.7, defining the value of the optimal information-based policy, is then

$$V_{\pi_{\mathbf{I}}^*}(\mathbf{I}_k) = \max_{\mathbf{u}_k} \left[R_{\mathbf{I}}(\mathbf{I}_k, \mathbf{u}_k) + \gamma \underset{\mathbf{I}_{k+1}}{E} \left[V_{\pi_{\mathbf{I}}^*}(\mathbf{I}_{k+1})|\mathbf{u}_k \right] \right] \tag{2.15}$$

$$= \max_{\mathbf{u}_k} \left[R_{\mathbf{I}}(\mathbf{I}_k, \mathbf{u}_k) + \gamma \underset{\mathbf{z}_{k+1}}{E} \left[V_{\pi_{\mathbf{I}}^*}(f_{\mathbf{I}}(\mathbf{I}_k, \mathbf{u}_k, \mathbf{z}_{k+1})) \right] \right] \tag{2.16}$$

The equivalence between MDPs and POMDPs viewed as I-state MDPs is summarised in Table 2.1. In principle, ordinary MDP solution methods can be used to solve Equation 2.16. The complication, as will be discussed in this chapter, is that information spaces are generally not discrete.

2.2.2 $\mathcal{I}_{hist}$ and Derived Information Spaces

An obvious way to describe the set of available information is with the history I-state $\mathbf{I}_{hist,k} \in \mathcal{I}_{hist,k}$, where $\mathbf{I}_{hist,k}$ is the vector of all available information:

$$\mathbf{I}_{hist,k} = (c_0, \mathbf{z}_0, \ldots, \mathbf{z}_k, \mathbf{u}_0, \mathbf{u}_1, \ldots, \mathbf{u}_{k-1}), \quad \mathbf{I}_{hist,0} = (c_0, \mathbf{z}_0) \tag{2.17}$$

The history I-space at time k, $\mathcal{I}_{hist,k}$, is therefore the space of all possible vectors $\mathbf{I}_{hist,k}$:

$$\mathcal{I}_{hist,k} = C_0 \times \underbrace{Z \times Z \cdots \times Z}_{k+1} \times \underbrace{U \times U \cdots \times U}_{k} \qquad (2.18)$$

For problems without a finite number of time intervals, $\mathcal{I}_{hist}$ is given by:

$$\mathcal{I}_{hist} = \mathcal{I}_{hist,0} \cup \mathcal{I}_{hist,1} \cup \mathcal{I}_{hist,2} \cup \cdots \qquad (2.19)$$

Operating on history I-states directly is clearly problematic, since the size of $\mathbf{I}_{hist,k}$ grows linearly with k. Acting using history I-states would involve the task of constructing a policy mapping from a history of arbitrary length to an action. Instead, it is more convenient to operate in a derived I-space, denoted $\mathcal{I}_{der}$, in which the available information can be represented more compactly [64].

Derived Information Spaces

The use of derived I-spaces greatly simplifies the task of constructing a policy. Rather than being a function of arbitrary length histories, the policy can be a function of finite-sized summaries of histories. Ideally the derived I-space should simplify the problem as much as possible, while retaining as much information as possible. These are often competing objectives since simpler I-spaces can result in information loss and, consequently, the inability to differentiate between distinct information histories. This is likely to cause a deterioration in the quality of plans.

To operate in a derived information space, an information mapping function is required [64][112][65]. Let $\kappa_{der} : \mathcal{I} \to \mathcal{I}_{der}$ denote an information mapping function, or I-map, which maps from an original information space to a derived information space. I-maps may be chained together to create new I-maps: given the I-maps $\kappa_{der} : \mathcal{I} \to \mathcal{I}_{der}$ and $\kappa_{der'} : \mathcal{I}_{der} \to \mathcal{I}_{der'}$, an I-state $\mathbf{I}$ can be transformed using $\mathbf{I}_{der'} = \kappa_{der'}(\kappa_{der}(\mathbf{I}))$.

Sufficient I-Maps

The information map κ_{der} is termed *sufficient* [64], given an I-state transition function $f_{\mathbf{I}}$, if a derived transition function $f_{\mathbf{I}_{der}}$ exists such that the following holds:

$$\forall \mathbf{I}_k \in \mathcal{I} : \kappa_{der}\big(f_{\mathbf{I}}(\mathbf{I}_k, \mathbf{u}_k, \mathbf{z}_{k+1})\big) = f_{\mathbf{I}_{der}}\big(\kappa_{der}(\mathbf{I}_k), \mathbf{u}_k, \mathbf{z}_{k+1}\big) \qquad (2.20)$$

This implies that for a given trajectory through I-space and a *sufficient* I-map, the same final state can be reached by either (a) evaluating the trajectory in $\mathcal{I}$ then mapping to $\mathcal{I}_{der}$, or (b) mapping to $\mathcal{I}_{der}$ initially then evaluating the trajectory in $\mathcal{I}_{der}$ using $f_{\mathbf{I}_{der}}$, as illustrated

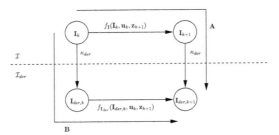

Figure 2.3: The I-map κ_{der} is termed *sufficient* if both path A (via $\mathbf{I}_{k+1}$) and B (via $\mathbf{I}_{der,k}$) produce the same derived state $\mathbf{I}_{der,k+1}$.

in Figure 2.3. If this is the case then the problem can be re-cast entirely in terms of derived I-states, and a solution sought in the simplified information space.

Insufficient I-Maps

In practice, problems are often re-cast in terms of derived I-spaces using insufficient I-maps, for which Equation 2.20 does not hold. While inexact results may be obtained by working in the derived information space, this is often outweighed by the benefits of simplicity.

When using an insufficient I-map, one approach to defining an approximate derived transition function $\hat{f}_{\mathbf{I}_{der}}$ is to use the transition function from the original space. Suppose a function $\hat{\kappa}_{der}^{-1}$ exists which maps from a point in the derived space back to a point in the original space. Then an approximate derived transition function can be constructed as

$$\hat{f}_{\mathbf{I}_{der}}(\mathbf{I}_{der,k}, \mathbf{u}_k, \mathbf{z}_{k+1}) = \kappa_{der}\left(f_{\mathbf{I}}\left(\hat{\kappa}_{der}^{-1}(\mathbf{I}_{der,k}), \mathbf{u}_k, \mathbf{z}_{k+1}\right)\right) \qquad (2.21)$$

This equation maps $\mathbf{I}_{der,k}$ back into the original I-space, applies the transition function $f_{\mathbf{I}}$ to produce $\mathbf{I}_{k+1}$, then maps $\mathbf{I}_{k+1}$ back into the derived I-space, as illustrated in Figure 2.4.

It remains to define $\hat{\kappa}_{der}^{-1}$. Let the pre-image of $\mathbf{I}_{der,k}$ refer to the set of points in $\mathcal{I}$ which κ_{der} maps to $\mathbf{I}_{der,k}$. For sufficient I-maps, the pre-image of $\mathbf{I}_{der,k}$ may be either a single point or a set of points. For insufficient I-maps, however, the pre-image is always a set of points[2]. Equation 2.21 can be implemented by defining $\hat{\kappa}_{der}^{-1}$ to simply select some point from the pre-image.

---

[2]If there were a one-to-one mapping between points in $\mathcal{I}$ and points in $\mathcal{I}_{der}$, Equation 2.21 would not involve an approximation and could be used to satisfy Equation 2.20.

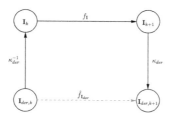

Figure 2.4: For an insufficient I-map, an approximate transition function $\hat{f}_{\mathbf{I}_{der}}$ can be derived from its counterpart in the original I-space.

Useful Information Spaces

Since solving POMDPs directly in $\mathcal{I}_{hist}$ is problematic, proposed POMDP solution methods generally operate in spaces derived from $\mathcal{I}_{hist}$. The choice of I-space has important implications for the type of solution algorithms which can be applied, the scalability of those algorithms, and the accuracy of the resulting plans.

The following sections describe a number of I-spaces which have been chosen, along with their I-maps and derived transition and reward functions. The spaces which will be described and the relationships between them are shown in Figure 2.5.

2.3 Belief States and $\mathcal{I}_{prob}$

An important I-space is $\mathcal{I}_{prob}$. κ_{prob} maps from an information history $\mathbf{I}_{hist,k}$ to a probability distribution over state-space $p(\mathbf{x}_k|\mathbf{I}_{hist,k})$, denoted $\mathbf{I}_{prob,k}$. The Markov assumption asserts that $p(\mathbf{x}_k|\mathbf{I}_{hist,k})$ is a sufficient statistic for an entire history, implying that

$$p(\mathbf{x}_{k+1}|\mathbf{I}_{hist,k}, \mathbf{u}_k, \mathbf{z}_{k+1}) = p(\mathbf{x}_{k+1}|\mathbf{I}_{prob,k}, \mathbf{u}_k, \mathbf{z}_{k+1}) \qquad (2.22)$$

and that κ_{prob} is a sufficient I-map.

$\mathbf{x}_k$	$\rightarrow$	$\mathbf{x}$
$\mathbf{x}_{k+1}$	$\rightarrow$	$\mathbf{x}^+$
$\mathbf{I}_k$	$\rightarrow$	$\mathbf{I}$
$\mathbf{I}_{k+1}$	$\rightarrow$	$\mathbf{I}^+$
$\mathbf{u}_k$	$\rightarrow$	$\mathbf{u}$
$\mathbf{z}_{k+1}$	$\rightarrow$	$\mathbf{z}^+$

Table 2.2: A summary of a simplified notation, omitting the subscript k and replacing the subscript $k+1$ with the superscript $+$.

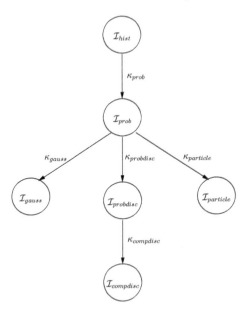

Figure 2.5: Important I-spaces for which solution algorithms have been proposed, and their relationships. $\mathcal{I}_{hist}$ is discussed in Section 2.2.2, $\mathcal{I}_{prob}$ in Section 2.3, and $\mathcal{I}_{probdisc}$ in Section 2.4, while $\mathcal{I}_{particle}$, $\mathcal{I}_{compdisc}$, and $\mathcal{I}_{gauss}$ are discussed in Section 2.5.

In order to simplify notation, we make the following substitutions from this point onwards. For quantities at the current time interval, the subscript k is omitted. For quantites at the next time interval, the subscript $k + 1$ is replaced with the superscript $+$. These replacements are summarized in Table 2.2. Hence Equation 2.22 becomes

$$p(\mathbf{x}^+|\mathbf{I}_{hist}, \mathbf{u}, \mathbf{z}^+) = p(\mathbf{x}^+|\mathbf{I}_{prob}, \mathbf{u}, \mathbf{z}^+) \tag{2.23}$$

The probability distribution $\mathbf{I}_{prob}$ is usually referred to as a belief, hence the words 'belief' and 'I-state' will be used interchangeably throughout this document to refer to I-states in $\mathcal{I}_{prob}$ and any of its derived spaces, relying on context to clarify ambiguities. By working in $\mathcal{I}_{prob}$, the POMDP over unknown states is transformed into an MDP over known beliefs. The agent's aim becomes that of selecting actions which coerce as much probability mass as possible towards high-reward states. Shifting probability mass can equally be viewed as shifting a point in $\mathcal{I}_{prob}$.

In order to operate in $\mathcal{I}_{prob}$, the I-state MDP models for rewards, transitions, and observations, referenced in Table 2.1, must be derived in terms of the underlying environment model. The reward model is simply an expectation over state-space:

$$R_{prob}(\mathbf{I}_{prob}, \mathbf{u}) = \underset{\mathbf{x}}{E}\big[R(\mathbf{x}, \mathbf{u})|\mathbf{I}_{prob}\big] \tag{2.24}$$

$$= \int_{\mathbf{x}} R(\mathbf{x}, \mathbf{u})\mathbf{I}_{prob}(\mathbf{x})d\mathbf{x} \tag{2.25}$$

Using $\mathbf{I}_{prob}^{-}$ to denote the belief after acting but before observing, given by

$$\mathbf{I}_{prob}^{-}(\mathbf{x}^{+}) = \int_{\mathbf{x}} p(\mathbf{x}^{+}|\mathbf{x}, \mathbf{u})\mathbf{I}_{prob}(\mathbf{x})d\mathbf{x} \tag{2.26}$$

the I-state transition function is

$$\mathbf{I}_{prob}^{+}(\mathbf{x}^{+}) = f_{\mathbf{I}_{prob}}(\mathbf{I}_{prob}, \mathbf{u}, \mathbf{z}^{+}) \tag{2.27}$$

$$= Cp(\mathbf{z}^{+}|\mathbf{x}^{+}, \mathbf{u})\mathbf{I}_{prob}^{-}(\mathbf{x}^{+}) \tag{2.28}$$

where C is a normalising constant which ensures that $\mathbf{I}_{prob}^{+}(\mathbf{x}^{+})$ integrates to one. The I-state transition function has a prediction-correction form familiar in robotics (*e.g.* [110]). Equation 2.26 projects $\mathbf{I}_{prob}$ forward according to the process model $p(\mathbf{x}^{+}|\mathbf{x}, \mathbf{u})$ to produce the prediction $\mathbf{I}_{prob}^{-}$, while Equation 2.28 uses Bayes' rule to correct that prediction using the observation likelihood function $p(\mathbf{z}^{+}|\mathbf{x}^{+}, \mathbf{u})$. The I-map $\kappa_{prob}(\mathbf{I}_{hist})$ can be evaluated using the initial conditions c_0 and repeated applications of $f_{\mathbf{I}_{prob}}$.

The I-state observation model is

$$p(\mathbf{z}^{+}|\mathbf{I}_{prob}, \mathbf{u}) = \int_{\mathbf{x}^{+}} p(\mathbf{z}^{+}|\mathbf{x}^{+}, \mathbf{u})\mathbf{I}_{prob}^{-}(\mathbf{x}^{+})d\mathbf{x}^{+} \tag{2.29}$$

Equation 2.29 may be unfamiliar to those with an estimation background. It represents the likelihood of observations in the immediate future given the current *belief* and action (as opposed to the current *state* and action). This quantity is required for planning but not for estimation: a planner needs to know the likelihood of future observations, whereas an estimator needs to know only how to incorporate them after they arrive.

Unfortunately, representing points in $\mathcal{I}_{prob}$ can be problematic. In the general continuous case, $\mathcal{I}_{prob}$ is the space of arbitrary continuous distributions over X and hence can be viewed as an infinite-dimensional vector space. To see why, consider approximating the function $p(\mathbf{x}|\mathbf{I}_{hist})$ with a series expansion. An infinite number of coefficients may be required to represent the distribution exactly [64].

In order to represent points in $\mathcal{I}_{prob}$ the space must first be transformed with further (probably insufficient) I-maps. To be useful, the transformations should allow distributions to be represented using finite vectors. If general distributions can occur, this necessarily involves reducing the set of distributions which can be represented. The following sections describe several possible mappings. Each mapping corresponds to a different approximation of the original continuous distributions, with different implications for the POMDP solution methods.

2.4   Discretised State Spaces: $\kappa_{probdisc}$

A well-studied approach to representing probability distributions over continuous states is to discretise the state-space, partitioning X into a set of cells $s \in S$. This approach, related to state aggregation in MDPs [99][114], can be viewed as the application of the I-map $\kappa_{probdisc}$ to $\mathcal{I}_{prob}$ to produce $\mathcal{I}_{probdisc}$, where $\mathcal{I}_{probdisc}$ is the space of fixed-length vectors of cell probabilities. $\kappa_{probdisc}$ transforms $\mathbf{I}_{prob}(\mathbf{x})$ by integrating the total probability mass in each discrete cell, producing $\mathbf{I}_{probdisc}(s)$. $\mathbf{I}_{probdisc}(s)$ is a vector of discrete cell probabilities of length $|S| - 1$, where the final cell probability is unnecessary due to the constraint that the probabilities sum to one. This is clearly not a sufficient I-map for continuous underlying state-spaces.

The discrete I-state reward, observation and transition functions are similar to their counterparts in $\mathcal{I}_{prob}$:

$$R_{probdisc}(\mathbf{I}_{probdisc}, \mathbf{u}) = \sum_{s} \mathbf{I}_{probdisc}(s) R_s(s, \mathbf{u}) \tag{2.30}$$

$$\mathbf{I}_{probdisc}^{-}(s^{+}) = \sum_{s} p(s^{+}|s, \mathbf{u}) \mathbf{I}_{probdisc}(s) \tag{2.31}$$

$$\mathbf{I}_{probdisc}^{+} = f_{\mathbf{I}_{probdisc}}(\mathbf{I}_{probdisc}, \mathbf{u}, \mathbf{z}^{+}) \tag{2.32}$$

$$= C p(\mathbf{z}^{+}|s^{+}, \mathbf{u}) \mathbf{I}_{probdisc}^{-}(s^{+}) \tag{2.33}$$

where C is a normalising constant which ensures that $\mathbf{I}_{probdisc}^{+}$ sums to one, and

$$p(\mathbf{z}^{+}|\mathbf{I}_{probdisc}, \mathbf{u}) = \sum_{s^{+}} p(\mathbf{z}^{+}|s^{+}, \mathbf{u}) \mathbf{I}_{probdisc}^{-}(s^{+}) \tag{2.34}$$

$R_s(s, \mathbf{u})$, $p(\mathbf{z}^{+}|s^{+}, \mathbf{u})$ and $p(s^{+}|s, \mathbf{u})$ refer to the discrete-state-based reward, observation and transition functions respectively. They can be determined using integrations over the original continuous space, assuming an inverse I-map $\hat{\kappa}_{probdisc}^{-1}$ exists to map from distributions over discrete cells back to continuous distributions. Appropriate choices may be to map to a mixture of uniform distributions over the areas defined by each cell, or to map to a set of delta functions centred on each cell. Chapter 3 will show an example of the former choice.

2.4.1 Gradient-Based Solution Methods

Since discrete state-spaces are not the focus of this book, the discussion of gradient-based methods is relatively brief. For a more detailed discussion, readers are referred to the references provided. For example, a good introduction is provided in [23].

While a discrete state-space means that beliefs can be represented using vectors of length $|S| - 1$, it remains to represent the value function over the resulting continuous belief-space. Many discrete POMDP solution methods rely on the result that the value function is piecewise-linear and convex (PWLC) [100][102]. Therefore the value function can be represented by the supremum of a finite set of n hyperplanes over the belief simplex:

$$V_\pi(\mathbf{I}_{probdisc}) = \max_i \mathbf{I}_{probdisc} \cdot \alpha_i \qquad (2.35)$$

for some set of α-vectors $\Gamma = \alpha_0, \alpha_1, ..., \alpha_i$ where α_i is a $|S| - 1$ dimensional vector specifying the hyperplane's axis-intercepts in belief-space. This result allows the value over a continuum of belief points to be represented exactly with the finite set of scalars describing Γ_k.

Assuming discrete action and observation spaces, a second important result is that if the value function at time $k + 1$ is a PWLC function, represented by the set Γ^+, then the value function at time k is also a PWLC function and can be represented exactly by a set of vectors Γ [102]. This update can be performed in two steps. The first generates intermediate sets of vectors $\Gamma^{\mathbf{u},\mathbf{z}^+}$, $\forall \mathbf{u} \in U$, $\forall \mathbf{z}^+ \in Z$:

$$\Gamma^{\mathbf{u},\mathbf{z}^+} \leftarrow \alpha_i^{\mathbf{u},\mathbf{z}^+}(s), \ \forall i \in \{1 \ldots |\Gamma^+|\} \qquad (2.36)$$

where

$$\alpha_i^{\mathbf{u},\mathbf{z}^+}(s) = \frac{1}{|Z|} R(s,\mathbf{u}) + \gamma \sum_{s^+ \in S} p(s^+|s,\mathbf{u}) p(\mathbf{z}^+|s^+,\mathbf{u}) \alpha_i^+(s^+) \qquad (2.37)$$

The second step generates a new value function by adding vectors α_i to Γ. Each α_i is obtained by choosing a particular action $\mathbf{u}'$, then selecting one $\alpha^{\mathbf{u}',\mathbf{z}^+}$ from each set $\Gamma^{\mathbf{u}',\mathbf{z}^+}$ and taking the sum:

$$\alpha_i = \sum_{\mathbf{z}^+ \in Z} \text{alphaselect}(\mathbf{u}', \mathbf{z}^+, i) \qquad (2.38)$$

where the operator alphaselect$(\mathbf{u}', \mathbf{z}^+, i)$ chooses the vector in $\Gamma^{\mathbf{u}',\mathbf{z}^+}$ used to create α_i.

Defining the cross-sum operator $\oplus$ as

$$\{a, b, \ldots\} \oplus \{p, q, \ldots\} = \{a + p, a + q, b + p, b + q, \ldots\} \qquad (2.39)$$

a simple approach is to enumerate the complete set of possibilities:

$$\Gamma_{complete} = \bigcup_{\mathbf{u}} \Gamma^{\mathbf{u}} \qquad (2.40)$$

where $\Gamma^{\mathbf{u}}$ is the cross-sum over observations:

$$\Gamma^{\mathbf{u}} = \Gamma^{\mathbf{u},\mathbf{z}_1^+} \oplus \Gamma^{\mathbf{u},\mathbf{z}_2^+} \oplus \cdots \tag{2.41}$$

In practice the complete set is unlikely to be necessary since many vectors will be dominated ($\forall \mathbf{I}_{probdisc} \, \exists j : \alpha_i \cdot \mathbf{I}_{probdisc} < \alpha_j \cdot \mathbf{I}_{probdisc}$) and hence not contribute to the value function. A number of algorithms perform exact value iteration by finding the minimal set of α-vectors required at each step, either by enumerating a superset of the required vectors and pruning the useless ones [73][36][25] or by iteratively expanding a subset until the minimal set has been found [102][27][60]. For a more thorough review of exact algorithms, the reader is directed to [23].

Finding the minimum set of required vectors is important because the size of Γ can grow rapidly. In the worst case, the number of vectors required at time k is given by

$$|\Gamma| = |U||\Gamma^+|^{|Z|}$$

where $|U|$, $|Z|$ and $|\Gamma^+|$ are the number of actions, observations, and vectors representing the value function at time $k+1$, respectively. Therefore the number of vectors required to represent the value function after i iterations is of the order $O(|U|^{|Z|^{i-1}})$ [53].

Unfortunately, strategies for finding the minimal set of vectors to represent the value function exactly are usually computationally expensive and seem to make a difference only in the constant factors rather than the order of the growth [86]. As a result, exact algorithms are generally considered to be intractable for all but trivial problems.

2.4.2 Approximate Gradient-Based Solutions: Point-Based Methods

Rather than generating all α-vectors required to represent the entire value function exactly, a number of algorithms perform approximate value iteration by generating only those vectors which maximise the value at a discrete set of belief points B [103][86][101][68][88][53].

Let backup($\mathbf{I}_{probdisc}$) denote the operator which returns the $\alpha \in \Gamma$ which maximises the value at belief point $\mathbf{I}_{probdisc}$ given the vectors Γ^+:

$$\text{backup}(\mathbf{I}_{probdisc}) = \underset{\{g_{\mathbf{u}}^{\mathbf{I}}\}_{\mathbf{u} \in U}}{\arg\max} \, \mathbf{I}_{probdisc} \cdot g_{\mathbf{u}}^{\mathbf{I}} \tag{2.42}$$

where

$$g_{\mathbf{u}}^{\mathbf{I}} = \sum_{\mathbf{z}} \underset{\alpha_i^{\mathbf{u},\mathbf{z}^+} \in \Gamma^{\mathbf{u},\mathbf{z}^+}}{\arg\max} \, \mathbf{I}_{probdisc} \cdot \alpha_i^{\mathbf{u},\mathbf{z}^+} \tag{2.43}$$

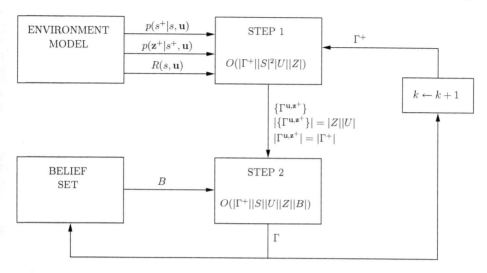

Figure 2.6: The structure of gradient-based POMDP solution methods, using a point-based approximation. Step 1 uses the environment model to generate intermediate sets of alpha vectors, $\{\Gamma^{\mathbf{u},\mathbf{z}^+}\}$, in time $O(|\Gamma^+||S|^2|U||Z|)$. Step 1 creates $|Z||U|$ intermediate sets, each containing $|\Gamma^+|$ vectors. Step 2 selects from and sums these intermediate vectors, given the belief set B, to produce the set of α-vectors for the next iteration. The time required for step 2 is $O(|\Gamma^+||S||U||Z||B|)$. Some algorithms also update the belief set during value iteration.

Point-based algorithms follow the two steps listed in the previous section. In step 1 the sets $\Gamma^{\mathbf{u},\mathbf{z}^+}$ are generated using Equations 2.36 and 2.37. In step 2, vectors are added to Γ using the backup operator and the belief set B, as illustrated in Figure 2.6. A common approach is to update B during planning, interleaving phases of value iteration with phases of belief set expansion [101][115][53][92][53][86]. Given the convexity of the value function, the subset of alpha-vectors induced by B defines a lower bound to that value function.

The advantage of point-based approaches is that they avoid the double-exponential growth in α-vectors experienced by exact algorithms. Instead, the maximum number of α-vectors (the size of the set Γ produced by step 2) is limited to $|B|$. The PERSEUS algorithm reduces the size of Γ further [103]. Since the gradient of each α-vector allows it to generalise over an area of the belief-space, a single vector may improve the value (but not necessarily provide the best value) at multiple belief points. This can be exploited in step 2 to produce a set Γ containing a number of vectors much less than $|B|$, while still ensuring that all belief points in B are improved. This has been shown to reduce computation time significantly.

The complexity of a single step of value iteration, using a point-based approximation, is [84]

$$O(|\Gamma^+||S|^2|U||Z| + |\Gamma^+||S||U||Z||B|) \tag{2.44}$$

In the worst case, $|\Gamma^+| = |B|$. The PERSEUS algorithm produces a set Γ^+ much smaller than $|B|$, but may require more iterations to converge due to the fact that belief points are not updated on every iteration. In Chapters 3 and 4 we compare directly against the PERSEUS algorithm.

Porta *et al.* generalise the idea of point-based updates to continuous state spaces, assuming discrete actions and observations [89]. The notion of α-vectors is generalised to α-functions (α-vectors over an infinite number of states), and sums over states are replaced with integrals. It is shown how these integrals can be evaluated in the case where the beliefs and action, observation, and reward models are mixtures of Gaussians over state-space. One complication is that the number of components in the mixtures representing beliefs and α-functions increases exponentially. To prevent this, the number of components in each function is kept constant by approximating the function, at each iteration, with a mixture of fewer components. While this is a promising approach, it is as yet unclear whether the computational cost of each update will scale beyond the simple one-dimensional problem presented, or whether the observation functions of realistic scenarios can be adequately described by a discrete set of mixtures of Gaussians over state-space.

Hoey *et al.* extend point-based value iteration to continuous observation spaces, using the fact that observations are useful only to the extent that they lead to different courses of action [56]. The observation space can therefore be partitioned by calculating the thresholds at which different observations require different actions. It is unclear how appropriate this is for robot navigation problems in which the action space is fundamentally continuous, and ideally every observation should lead to a different action.

The SPOVA algorithm uses a smooth, continuous function to approximate the set of hyperplanes which maximise the value function at a set of belief points [82]. By using a differentiable function, the error between the smooth approximation and the true value function can be minimised using gradient descent.

2.5 Beyond $\mathcal{I}_{probdisc}$

There are many possibilities besides $\mathcal{I}_{probdisc}$ for representing continuous beliefs. The major advantage of $\mathcal{I}_{probdisc}$ is the fact that the value function is PWLC, allowing gradient-based value iteration. For many of the other I-spaces shown in Figure 2.5, the value function is in general not PWLC and hence gradient-based value iteration is not possible. An algorithm that is feasible for all I-spaces in Figure 2.5, including $\mathcal{I}_{probdisc}$, is fitted value iteration (FVI) [49]. Since FVI is relied upon heavily in this document, Sections 2.5.1 and 2.5.2 describe the algorithm. Section 2.5.3 then discusses the application of FVI to the various I-spaces shown in Figure 2.5.

2.5.1 Fitted Value Iteration

Fitted Value Iteration (FVI) is an approach to solving MDPs with large or infinite numbers of states. As shown in Sections 2.2 and 2.3, a POMDP can be converted to a continuous belief-state MDP. FVI can therefore be applied to this resultant (infinite-state) MDP.

The central idea behind FVI is to store values explicitly at only a relatively small number of states, using a function approximator to approximate the value function for all states in between. In principle the value at one state may confer no information about the value at another state. However if the value function is sufficiently smooth and enough values are stored explicitly, FVI is likely to provide a good approximation. At each new time-step, a new set of explicit values can be estimated from the approximate value function of the old time-step.

More formally, let G be a set of states of size $|G|$, $G = \{\mathbf{x}_{G,1}, \mathbf{x}_{G,2}, \dots, \mathbf{x}_{G,|G|}\}$, and let Ψ_G be the set of explicit state-value pairs

$$\Psi_G = \left\{ \left(\mathbf{x}_{G,1}, \psi(\mathbf{x}_{G,1}) \right), \left(\mathbf{x}_{G,2}, \psi(\mathbf{x}_{G,2}) \right), \dots, \left(\mathbf{x}_{G,|G|}, \psi(\mathbf{x}_{G,|G|}) \right) \right\} \tag{2.45}$$

where $\psi(\mathbf{x}_{G,i})$ is the estimated value of the i'th state in G. Let $\hat{V}(\mathbf{x})$ denote the current estimated value of any state $\mathbf{x} \in X$. $\hat{V}(\mathbf{x})$ can be estimated using a function approximator ϕ_G, based on the set Ψ_G

$$\hat{V}(\mathbf{x}) = \phi_G(\mathbf{x}, \Psi_G) \tag{2.46}$$

The value of the i'th state in G can then be estimated at time k from the approximate value function at time $k+1$ by replacing the true value function in equation 2.7 with its approximate version

$$\psi(\mathbf{x}_{G,i}) = \max_{\mathbf{u}} \left[R(\mathbf{x}_{G,i}, \mathbf{u}) + \gamma \underset{\mathbf{x}^+}{E} \left[\hat{V}^+(\mathbf{x}^+)|\mathbf{u} \right] \right] \tag{2.47}$$

For clarity, the operation of Equation 2.47 is depicted in Figure 2.7.

Fitted value iteration for a discounted MDP is guaranteed to converge provided the function approximator is not an expansion in the max norm [49]. This is the case for convex function approximators. Loosely speaking, a convex function approximator is one which estimates the value of a state as a weighted sum of the values of nearby states.

More formally, let $\lambda_G(\mathbf{x}, j)$ denote a weighting function defined for the set G, which takes an arbitrary state $\mathbf{x}$ and the index j of a state in G, and returns a weighting. Then a convex

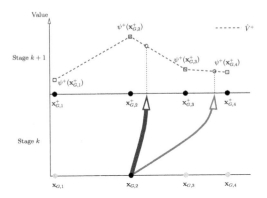

Figure 2.7: An example demonstrating fitted value iteration in a one-dimensional state space, using linear interpolation as the function approximator ϕ_G. The function approximation is shown as a dashed line, interpolating between the point-value pairs $\left(\mathbf{x}_{G,i}^+, \psi^+(\mathbf{x}_{G,i}^+)\right)$ to produce $\hat{V}^+$. The value of $\mathbf{x}_{G,2}$, denoted $\psi(\mathbf{x}_{G,2})$, can be calculated using Equation 2.47. The diagram shows the calculation of the expectation term $\underset{\mathbf{x}^+}{E}\left[\hat{V}^+(\mathbf{x}^+)|\mathbf{u}\right]$. For the action $\mathbf{u}$, the two possible state transitions are indicated by the two arrows. The thickness of each arrow corresponds to the probability of the transition. $\underset{\mathbf{x}^+}{E}\left[\hat{V}^+(\mathbf{x}^+)|\mathbf{u}\right]$ is equal to the sum of $\hat{V}^+$ at the arrow terminations, weighted by the transition probabilities.

function approximation rule is given by

$$\phi_G(\mathbf{x}, \Psi_G) = \sum_{j=1}^{|G|} \lambda_G(\mathbf{x}, j)\psi_k(\mathbf{x}_{G,j}) \tag{2.48}$$

where $0 \leq \lambda_G(\mathbf{x}, j) \leq 1 \ \forall j$ and $\sum_{j=1}^{|G|} \lambda_G(\mathbf{x}, j) = 1$ [53]. Fortunately, the family of convex rules includes many commonly-used rules such as nearest neighbour, kernel regression, linear point interpolation and others. The remainder of this document assumes the use of convex function approximators, therefore it is assumed that a function approximator is fully specified by a weighting function.

2.5.2 Converting a Continuous MDP to a Discrete MDP of Size $|G|$

It can be shown that, for a fixed set of discrete actions and a convex function approximator, FVI can be used to convert a large or continuous MDP to a discrete MDP with a number of states equal to $|G|$ [49][53].

Equation 2.47 can be written

$$\psi_k(\mathbf{x}_{G,i}) = \max_{\mathbf{u}} \left[R(\mathbf{x}_{G,i}, \mathbf{u}) + \gamma \int_{\mathbf{w}^+} p(\mathbf{w}^+|\mathbf{x}_{G,i}, \mathbf{u}) \hat{V}^+(\mathbf{x}') d\mathbf{w}^+ \right] \tag{2.49}$$

where $\mathbf{x}' = f(\mathbf{x}_{G,i}, \mathbf{u}, \mathbf{w}^+)$. Substituting the definition of the estimated value function from Equation 2.46 gives

$$\psi_k(\mathbf{x}_{G,i}) = \max_{\mathbf{u}} \left[R(\mathbf{x}_{G,i}, \mathbf{u}) + \gamma \int_{\mathbf{w}^+} p(\mathbf{w}^+|\mathbf{x}_{G,i}, \mathbf{u}) \phi_G(\mathbf{x}', \Psi_G^+) d\mathbf{w}^+ \right] \tag{2.50}$$

Using Equation 2.48, the convexity of the function approximator allows Equation 2.50 to be written as

$$\psi(\mathbf{x}_{G,i}) = \max_{\mathbf{u}} \left[R(\mathbf{x}_{G,i}, \mathbf{u}) + \gamma \int_{\mathbf{w}^+} \sum_{j=1}^{|G|} p(\mathbf{w}^+|\mathbf{x}_{G,i}, \mathbf{u}) \lambda_G(\mathbf{x}', j) \psi^+(\mathbf{x}_{G,j}^+) d\mathbf{w}^+ \right] \tag{2.51}$$

$$= \max_{\mathbf{u}} \left[R(\mathbf{x}_{G,i}, \mathbf{u}) + \gamma \sum_{j=1}^{|G|} \psi^+(\mathbf{x}_{G,j}^+) \int_{\mathbf{w}^+} p(\mathbf{w}^+|\mathbf{x}_{G,i}, \mathbf{u}) \lambda_G(\mathbf{x}', j) d\mathbf{w}^+ \right] \tag{2.52}$$

$$= \max_{\mathbf{u}} \left[R(\mathbf{x}_{G,i}, \mathbf{u}) + \gamma \sum_{j=1}^{|G|} \psi^+(\mathbf{x}_{G,j}^+) T(\mathbf{x}_{G,i}, \mathbf{u}, \mathbf{x}_{G,j}^+) \right] \tag{2.53}$$

where

$$T(\mathbf{x}_{G,i}, \mathbf{u}, \mathbf{x}_{G,j}^+) = \int_{\mathbf{w}^+} p(\mathbf{w}^+|\mathbf{x}_{G,i}, \mathbf{u}) \lambda_G(f(\mathbf{x}_{G,i}, \mathbf{u}, \mathbf{w}^+), j) d\mathbf{w}^+ \tag{2.54}$$

in the last step. For a fixed and discrete set of disturbances, the integral becomes a summation

$$T(\mathbf{x}_{G,i}, \mathbf{u}, \mathbf{x}_{G,j}^+) = \sum_{\mathbf{w}^+} p(\mathbf{w}^+|\mathbf{x}_{G,i}, \mathbf{u}) \lambda_G(f(\mathbf{x}_{G,i}, \mathbf{u}, \mathbf{w}^+), j) \tag{2.55}$$

$T(\mathbf{x}_{G,i}, \mathbf{u}, \mathbf{x}_{G,j}^+)$ can be interpreted as representing a probability, since $0 \leq T(\mathbf{x}_{G,i}, \mathbf{u}, \mathbf{x}_{G,j}^+) \leq 1$, and $\sum_{j=1}^{|G|} T(\mathbf{x}_{G,i}, \mathbf{u}, \mathbf{x}_{G,j}^+) = 1$ hold (given the definition of convexity and the fact that the disturbance probabilities sum to one). To clarify again, Figure 2.8 depicts the example from Figure 2.7 after conversion to a discrete MDP of size $|G|$.

It can be seen from Equation 2.53 that the set of transition probabilities can all be pre-calculated before value iteration begins, and stored as a matrix T, rather than being re-calculated at every iteration. This is a considerable computational saving: T contains $O(|G|^2|U|)$ non-zero entries, each of which requires forward simulation of the environment and an application of the weighting function.

After pre-calculation of T, Equation 2.53 is clearly very similar to the discrete version of the Bellman equation given in Equation 2.8. The explicit storage of the value of each state in Ψ_G implies that Algorithm 1 can be applied directly to solve the new MDP.

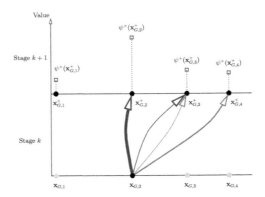

Figure 2.8: The fitted value iteration example from Figure 2.7, after conversion to a $|G|$-state discrete MDP. The thickness of the lines transitioning from stage k to $k+1$ are proportional to $T(\mathbf{x}_{G,i}, \mathbf{u}, \mathbf{x}_{G,j}^+)$. Note that $\hat{V}^+$, the value function estimate obtained through function approximation, is no longer required.

2.5.3 Application to Various I-Spaces

While they are not all presented as such, we would argue that many algorithms from the POMDP literature can be viewed as the application of fitted value iteration to a particular I-space, either with or without conversion to a discrete MDP. Under the view of a POMDP as an I-state MDP, it is relatively straightforward to apply FVI after making the substitutions listed in Table 2.1, and replacing the state set G with the belief set B. The mechanics are described in more detail in Chapter 3.

Fitted Value Iteration in $\mathcal{I}_{probdisc}$

Fitted value iteration directly in $\mathcal{I}_{probdisc}$, also known as a grid-based approximation, is well studied [68][126][53][14][17]. The main problem is the high dimensionality of the belief space. In general, the number of samples required to achieve a given sample density is exponential in the dimensionality of the space. When beliefs are represented as vectors of cell probabilities, the dimensionality of the belief space increases rapidly with both the size of the state space and the resolution of the discretisation.

Fitted Value Iteration in $\mathcal{I}_{particle}$

Thrun applies the idea of fitted value iteration to continuous state spaces [108]. This can be viewed as applying the I-map $\kappa_{particle}$ which maps from a point in $\mathcal{I}_{prob}$ by sampling from that distribution. The I-map is insufficient for a finite number of particles. Monte Carlo methods are used to evaluate the integrals governing the dynamics of the derived I-space.

The weighting function in belief-space is k-nearest-neighbour, where the distance metric is the KL divergence between beliefs. Since beliefs are represented by particle sets, a first step in calculating the KL divergence is to estimate the continuous belief distribution using kernel methods. The integration required for the KL divergence calculation is performed using Monte Carlo methods. Due to the computational costs involved in calculating the belief-space transition function and weighting function, it seems unlikely that this approach will scale to very large problems.

Fitted Value Iteration in $\mathcal{I}_{compdisc}$

The problems of the dimensionality of $\mathcal{I}_{probdisc}$ can be avoided by using dimensionality reduction techniques to compress the belief space. This can be viewed as employing the I-map $\kappa_{compdisc}$. Poupart *et al.* experiment with linear compression algorithms [90]. They note that linear compression is unlikely to be lossless and present an algorithm for finding the linear compression technique which minimises the reconstruction error. The I-map is sufficient only in the case of lossless compression.

This idea has been extended to non-linear compression algorithms [92][66]. The non-linearity of the compression algorithm breaks the convexity of the value function, and hence fitted value iteration is used as a solution method. Lacking an obvious low-dimensional model, the transition function must be calculated by mapping back and forth between $\mathcal{I}_{compdisc}$ and $\mathcal{I}_{probdisc}$ using equation 2.21.

Roy's AMDP algorithm [92] bears a strong relation to the algorithm presented in this book. AMDP compresses distributions by representing them using their mean and entropy. In this case, $\hat{\kappa}_{compdisc}^{-1}$ maps to the closest Gaussian distribution over discrete states. The important difference between AMDP and the work described in this book is that AMDP relies on an underlying discretisation. It will be shown in Section 4.5 that this difference has important scalability implications.

Fitted Value Iteration in $\mathcal{I}_{gauss}$

The I-space explored in this book is $\mathcal{I}_{gauss}$: the space of Gaussian approximations to continuous distributions. Lavalle suggests the approach of planning in the space of Gaussian approximations [65][64] but does not, to the author's knowledge, develop the idea further.

Continuous worlds with linear dynamics, quadratic costs, and Gaussian noise (so-called LQG problems) present a special case for $\mathcal{I}_{gauss}$. Under these conditions, the distribution over Gaussian posterior beliefs is itself a Gaussian. FVI need not be applied because a closed-form solution to value iteration exists [10]. Such environments are not particularly interesting from a planning perspective, however. They are said to be separable: given the estimate from a Kalman filter (which is optimal for linear worlds with Gaussian noise), the optimal plan is always to minimise the cost for the mean of the belief under the assumption of perfect state information [10].

In contrast, planning problems such as the ones considered in this book are interesting because they exhibit many sources of non-linearities and may have more complicated reward functions. In general the distribution over posterior beliefs is not Gaussian, and where the optimal plan in a linear world is to move directly towards the goal, such an approach is generally inapplicable in robot navigation planning problems.

2.6   Belief Set Selection

For both point-based gradient methods and fitted value iteration methods, an important step is to select a set of belief points B at which to update the value function. As discussed in Chapter 1, only a subset of the possible belief-space is likely to be encountered in practice. While restricting the class of allowable beliefs can help, the probability of reaching beliefs within that class is unlikely to be uniform over the entire I-space.

In the context of fitted value iteration, a trade-off exists between the computational cost of applying a weighting function and freedom in choosing beliefs. A simple approach is to sample the belief space on a regular grid [68]. While this approach does not focus computation on likely areas of the belief space, it allows for the use of an extremely fast linear weighting function based on Coxeter-Freudenthal-Kuhn triangulation [74] (or simply Freudenthal triangulation). This finds an interpolation based on $d + 1$ points rather than the 2^d points in the bounding hypercube, where d is the dimension of the space [31]. Freudenthal triangulations are described in more detail in Section 3.3.2. Using a variable-resolution grid allows a higher sample density in important areas of the belief space while still allowing a fast linear interpolation scheme [126][75]. Arbitrary point sets have also been used in the context of fitted value iteration [108][53][92], at the cost of a more expensive weighting function.

A number of strategies have been investigated for selecting arbitrary belief sets. Strategies for selecting B prior to planning include random selection, heuristics such as inclusion of the corners of the belief simplex [53][88], and forward-simulation of the model using either random actions [103] or actions from a heuristic policy [92]. Many algorithms also update B during planning, based on the uncertainty of the value function [101][115], high-value regions of the belief space [53], the policy generated using the current value function [92], random policies [53] or a policy designed to explore the belief space [86].

2.7 Non-Value-Iteration-Based Approaches

While this book focusses on POMDP solution methods based on value iteration, a number of other methods have been proposed. This section reviews heuristics, policy iteration, forward search, hierarchical approaches, and solution methods based on a view of POMDPs as graphical models.

2.7.1 Heuristic Approaches

Since POMDPs are difficult to solve, the most common approach by far (at least for real-time applications such as mobile robot navigation) is to use a heuristic approach to planning rather than the full POMDP solution. Heuristic approaches can be divided into three categories: (a) those which do not consider uncertainty at all, (b) MDP-based heuristics which consider stochastic actions but not future uncertainty, and (c) those which can act in order to resolve uncertainty. While this section provides a brief overview, more details are available in [24], [53], and the references therein.

Heuristics Without Uncertainty Considerations

Replan is a simple strategy, but probably the most widely used in practice. It simply plans under the assumption that the most likely state is true, and that the world is deterministic. If, during plan execution, the most likely state drifts far enough from the plan, it generates a new plan. We compare against *Replan* in the real robot navigation problem presented in Chapter 8.

MDP-Based Heuristics

For a sufficiently small discrete state-space, the solution to the underlying MDP is relatively easy to obtain. Two common heuristics based on the MDP solution are MLS and Q_{MDP}.

MLS, or *Most Likely State*, simply assumes that the most likely state is in fact the true state, and takes the corresponding action from the MDP policy [79]. This is a good approximation to the full POMDP solution when distributions are compact, and the most likely state is never far from the truth. We compare against MLS in subsequent chapters.

Q_{MDP} requires the entire MDP value function, and can be viewed as a voting system [67]. Given a belief over discrete states, each state votes on actions. The number of votes a state s_i can cast is proportional to the probability that s_i is the true state. s_i casts its allotted votes by voting on actions in proportion to their MDP value from state s_i. After voting, the agent takes the action with the most votes.

Q_{MDP} effectively assumes that all uncertainty will disappear after it takes its action. Indeed, it would be optimal if this assumption were true [24]. It can fail however when uncertainty is large, and unlikely to disappear after a single action. In highly uncertain scenarios, Q_{MDP} will take an action that is reasonable (in terms of reward) in most states. If this action does not resolve uncertainty, Q_{MDP} will continue to take it forever.

Heuristics Which Can Act to Resolve Uncertainty

The problem with heuristics discussed so far is that they only ever act to seek reward, possibly taking into consideration their current uncertainty and the uncertainty of their actions. Unfortunately, they will never act to decrease their future uncertainty. This kind of behaviour can be extremely important for an agent which is to operate robustly.

Action entropy is an example of a heuristic which can act to reduce uncertainty [24]. It switches between two distinct modes: seeking reward and seeking information. Recognising that uncertainty is problematic only when it introduces uncertainty about the appropriate action, *action entropy* uses the belief-optimality distribution as its switching criterion. When the entropy of this distribution is above a threshold, and therefore the agent is uncertain which action to take, *action entropy* takes the action which will best reduce its belief uncertainty over a one-step horizon. At other times, it follows one of the MDP-based heuristics.

Coastal navigation plans a fixed path, but considers the quality of localisation along that path [93]. It begins by calculating the information content of each state, based on the extent to which an observation from that state would modify a fixed prior. It then assigns a cost to each state as a weighted sum of the information-based cost and a goal-related cost.

These uncertainty-aware heuristics can be an improvement over simpler heuristics, but have shortcomings. Firstly, they are unable to make longer-term plans, reasoning about how uncertainty will evolve over the course of a plan. Secondly, they rely on a human designer to decide on the importance of certainty. This is a difficult parameter to specify, especially because it is not constant for a given problem or environment. Sometimes uncertainty is not problematic: uncertainty is undesirable if and only if it prevents an agent from achieving its aim. Similarly, sometimes an agent may be forced to persist with a high level of uncertainty, in a portion of the belief-space in which uncertainty-reducing actions are ineffective. In contrast, the full POMDP solution provides the optimal balance, seamlessly integrating information gathering and goal-directed behaviour, and reasoning about belief propagation over a significant time horizon.

2.7.2   Policy Iteration

This document focusses on value-based approaches, which attempt to find a value function over belief-space, from which a policy can be extracted. Rather than representing policies implicitly with a value function, an alternative is to represent policies explicitly and search the policy space directly. Given a representation for policies, policy iteration alternates between evaluating a candidate policy and producing a new candidate by modifying that policy. There is an equivalence between the two approaches: where value iteration extracts a policy from a converged value function, policy iteration calculates a value function from a policy during each policy evaluation step.

Hansen shows how policies can be represented as finite state machines (FSMs) [52]. Each node in the FSM dictates a particular action, while each arc corresponds to a particular observation. Each step of policy improvement involves modifying the FSM by adding and removing nodes, and changing the actions associated with nodes (which changes the successor nodes associated with observations). Modifications are based on exact updates, and hence convergence is guaranteed. Compared to exact value iteration, results show that this approach converges in fewer iterations. However, as with exact value iteration, it fails to scale to problems with more than a handful of states.

To improve scalability, a number of approximate approaches search for good policies within some restricted class. By selecting a smoothly parameterised policy class, gradient-based policy search can be applied directly to problems with continuous state-spaces, but can suffer from problems of local optima and low-gradient plateaus [76][124][8][77][61][1][72]. Ng *et al.* suggest the use of reward shaping for escaping low-gradient plateaus [76], however this requires the application of some domain knowledge. Bounded Policy Iteration [91] utilises a strategy for escaping local optima, while keeping policies simple. It uses gradient ascent to optimise policies represented as FSMs of a fixed size. When a local optimum is detected, extra nodes are added to allow the controller to break out of that optimum. Belief-based Stochastic Local Search [18] proposes another method for avoiding local optima. It alternates between gradient-based optimisation of a fixed-size FSM and FSM expansion by adding nodes corresponding to good but potentially un-reachable beliefs.

Policy iteration has shown strong promise. Its direct applicability to problems with large and continuous state spaces has made it successful in real-world applications such as helicopter control [6]. For the kinds of problems considered in this book, however, it is unclear whether a relatively simple controller will be capable of making the long term plans required for robot navigation.

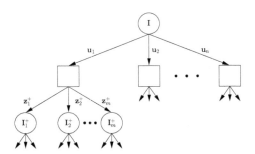

Figure 2.9: A POMDP viewed as a game-tree, starting from belief **I**. Circles represent nodes from which the agent chooses a value-maximising action from one of $n = |U|$ choices. Squares represent nodes from which the environment probabilistically chooses one of $m = |Z|$ observations. The value of each node is based on the rewards associated with belief-action transitions and the estimated values of the unexpanded leaf nodes.

2.7.3   Forward Search

The value-iteration-based approaches discussed so far work backwards in time. Each iteration assumes a value function estimate in the future. This is used to create a value function estimate for the present. Since this approach is generally both time-consuming and valid for the entire belief-space, it lends itself to offline computation.

In contrast, a number of POMDP solution algorithms search forwards in time, starting from the current belief [47][15][123][32][71][62][81]. The POMDP can be viewed as a game which alternates between the agent selecting an action and nature selecting an observation. A traditional approach to maximising performance in turn-based games is to represent the game as a tree [95]. Figure 2.9 depicts such a tree applied to the POMDP problem: circles represent nodes from which the agent selects an action, and squares represent nodes from which nature probabilistically selects an observation. The value of an action node involves a maximisation over the values of its children, where the value of an observation node involves an expectation. Given a heuristic to estimate the values of the unexpanded leaf nodes of the tree, a naive approach to solving the POMDP is brute-force search of this tree, expanding every action in a breadth-first order. Since the game-tree is valid only for the set of reachable beliefs from a known starting belief, this approach lends itself to online computation.

Relative to methods based on value iteration, forward search has a number of advantages. It requires no offline processing step, and can adapt to changes in the environment so long as the model is also updated [81]. For discrete actions and observations, a continuous belief-space does not present any particular problem because a value function needn't be represented over the entire belief space, but rather the set of reachable beliefs.

The limiting factor for forward search is that, at least for naive breadth-first search, the complexity is exponential in the planning horizon. More specifically, it scales with $(|U||Z|)^d$ where d is the depth to which the tree is searched. This approach is therefore unlikely to scale to problems which require an agent to make long-term plans. Kearns *et al.* reduce the dependence on $|Z|$ by calculating the expectation over observations by sampling rather than complete enumeration [62]. The computational complexity can be further reduced by scaling down the number of samples for calculations further down the tree, which have less effect on the topmost values due to the discount factor γ.

A number of authors reduce the computational complexity by expanding actions in a more appropriate order, using a search algorithm such as AO^* [78]. For AO^* to be effective, a good heuristic is required to estimate the value of un-expanded nodes. Example heuristics include problem-specific heuristics [81], and heuristics based on the solution of the underlying MDP [122][123][15].

An interesting approach to developing a heuristic is to use value iteration. Davies *et al.* calculate a coarse value function for the entire belief-space using value iteration [32]. This is then used to guide forward search, which refines the value function for the set of reachable beliefs. While the coarse value function may not provide AO^* with an *admissible* heuristic (*i.e.* one which always over-estimates the value), Davies *et al.* point out that an accurate but inadmissible heuristic is likely to provide better performance than an admissible heuristic with very loose bounds. A related approach is real-time dynamic programming, which amends the value function online, based on received rewards [15][47].

While this review has focussed on game-tree search for POMDPs, a more general and detailed review of game-tree search in AI problems is presented in Chapter 7. Chapter 7 also shows how forward planning can be incorporated into the POMDP solution method advocated in this book, and presents experimental results evaluating its effectiveness.

2.7.4 Hierarchical Approaches

Hierarchical approaches aim to decrease computational requirements by decomposing a large POMDP into a set of sub-POMDPs. The cost of solving the constituent sub-POMDPs can be significantly less than the cost of solving the original.

Theocharous proposes a hierarchical model for robot navigation in an office environment [107]. A set of abstract states are posited, each of which encapsulates a set of underlying states. A macro-action from an abstract state is equivalent to a set of actions through the underlying concrete state-space. Theocharous shows that the entropy of beliefs over abstract states is significantly less than the entropy over concrete states. Therefore the assumption of complete observability of abstract states, and the application of heuristics based on this assumption, is a better approximation than for concrete states.

While experiments show that planning is simplified by this hierarchical approach, Theocharous's environment is highly structured, consisting of corridors and junctions such that the interfaces between abstract states are tightly constrained. It is unclear how well the approach will generalise to problems which exhibit less structure.

Rather than specifying a hierarchy of states, Pineau specifies a hierarchy of actions [87]. A set of abstract actions are posited, each of which consists of a number of sub-tasks. This approach is much more applicable to problems involving discrete sets of actions; it is unclear how to build such a hierarchy for the kinds of robot navigation problems which are the subject of this book.

Foka specifies a hierarchy of both states and actions for robot navigation problem [41]. The hierarchy of states is reminiscent of a quad-tree decomposition [96]. The discretisation of both states and actions is finer at levels deeper in the hierarchy. Individual sub-POMDPs are solved using an MDP-based heuristic. While extensive results of the computational requirements are presented, the effects on performance are less clear.

The most serious problem limiting the application of hierarchical approaches is the requirement that the hierarchy be specified by a human designer, based on the perceived structure of the particular domain. A method for automating this process would be extremely valuable.

2.7.5   POMDPs as Graphical Models

POMDPs are often described using graphical models [83], as shown in Figure 2.10. At step k, the agent has access to the information $\mathbf{I}$ and must select an action $\mathbf{u}$. Although the next I-state $\mathbf{I}^+$ is a deterministic function of $\mathbf{I}$, $\mathbf{u}$ and $\mathbf{z}^+$, knowledge of only $\mathbf{I}$ and $\mathbf{u}$ induces the probability distribution over possible next I-states $p(\mathbf{I}^+|\mathbf{I}, \mathbf{u})$. The agent must make a decision based on this probability distribution over future I-states. After the decision has been made, an observation is revealed and a single I-state is selected from the distribution.

Inference in graphical models is the process of fixing certain nodes (usually the observable variables), then applying well-known inference algorithms to determine distributions over variables of interest. Attias proposes a novel approach to using general graphical model theory for solving POMDPs [5]. Actions are treated as random variables. If episodes are of a maximum length of N time-steps, the N'th state is fixed to be the goal state, and the first observation is fixed. Assuming a prior distribution over the (assumed random) action variables, standard inference algorithms can then be applied to find the Maximum A Priori (MAP) sequence of actions. Extensions are also suggested for incorporating general reward functions rather than assuming a single goal state. The central insight is that by casting the problem as inference in a graphical model, powerful general inference algorithms can be brought to bear.

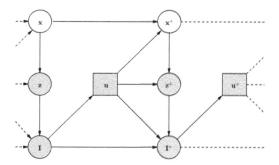

Figure 2.10: Two time slices of the POMDP problem, represented as a graphical model. Shaded nodes are observable, while the un-shaded state nodes are hidden. Rewards are omitted for clarity. Representing the POMDP as a graphical model shows how a joint probability distribution over all states, actions, observations, and I-states can be factored into smaller conditional probability distributions.

Verma and Rao extend this idea, noting that inference can be performed more efficiently by computing the MAP sequence of actions, states, and observations, rather than simply actions [119]. As the computed sequence of actions is followed, the occurrence of an unexpected observation causes the agent to re-plan. While the approach has shown promising results for small toy worlds, the cost of re-planning online may impede its application to real-world problems.

2.8   Summary

This chapter introduced the basic concepts and terminology required to discuss problems involving decision making under partial observability. Other than in very special cases, acting in continuous partially observable domains requires that a set of approximations be made in order to simplify the problem to the point where a tractable solution is available. A number of such simplifying assumptions were reviewed, along with the algorithms for solving the resultant problems.

Despite the rich set of solution methods available, POMDP algorithms applied to robot navigation problems have generally failed to scale beyond fairly unrealistic scenarios. The following chapter introduces a solution algorithm based on the simplifying assumption that belief distributions can be well approximated by Gaussians. We show that this is usually a reasonable approximation for continuous navigation problems and, with the addition of the improvements described in subsequent chapters, is capable of scaling to real-world problems.

Chapter 3

Parametric Information Spaces

The previous chapter reviewed related approaches, and laid the groundwork for the Parametric POMDP algorithm to be defined. We use PPOMDP to refer to a POMDP solution algorithm which maps from $\mathcal{I}_{prob}$ to an I-space in which continuous distributions can be represented with finite-length parameter vectors. The resultant continuous I-state MDP can be solved using fitted value iteration. This chapter discusses the general methodology in detail. Given a model of the world, the following steps are required:

1. Define the continuous I-state MDP:

    (a) Choose a parametric representation (*i.e.* an I-space $\mathcal{I}$)

    (b) Define the models in that I-space:

    - reward: $r_\mathbf{I} = R_\mathbf{I}(\mathbf{I}, \mathbf{u})$
    - transition: $\mathbf{I}^+ = f_\mathbf{I}(\mathbf{I}, \mathbf{u}, \mathbf{z}^+)$
    - observation: $p(\mathbf{z}^+ | \mathbf{I}, \mathbf{u})$

2. Use FVI to discretise the resultant I-state MDP:

    - Choose a belief set B and weighting function λ_B
    - Discretise the continuous I-state MDP, producing a discrete transition function T and reward function R.

3. Solve the discrete I-state MDP.

Figure 3.1 illustrates the approach and the relationships between the steps.

While the methodology as described is relatively straightforward, the accuracy of the resultant algorithm and its viability for application to real problems hinges on the particular choices made for each of the items above. As pointed out in Chapter 2, despite not all being presented as such,

43

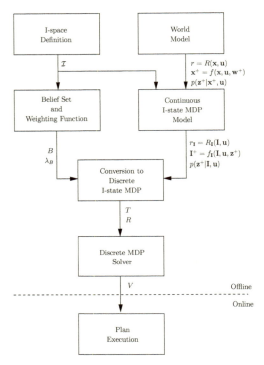

Figure 3.1: Overview of the relationships between the steps required to define, solve and execute the PPOMDP algorithm. The outputs of each step are shown. Note that extra arrows could be added: a possibility which is excluded here for simplicity is to interleave MDP solving with belief set expansion, as described in Chapter 2.

a number of algorithms from the literature are specific instances of this general methodology, or at least close variants thereof, but each has very different properties. The aim of this Chapter is to discuss some of the motivating factors for each step in Figure 3.1, and to make and evaluate some simple choices. The sum of these simple choices defines the Basic-PPOMDP algorithm, which provides a baseline upon which the remaining chapters of this book can improve. Each of the individual pieces which make up Basic-PPOMDP has at least been suggested in the literature, although perhaps not in the context of this methodology. Rather, the contribution of this chapter is to assemble several such pieces in a novel configuration, to experimentally validate the result, and to compare it against the state of the art.

Section 3.1 discusses factors influencing the choice of I-space. Section 3.2 argues for the use of $\mathcal{I}_{gauss}$, explaining why this choice is likely to be successful for robot navigation problems. Section 3.3 describes the choice of belief set and a weighting function based on a Freudenthal triangulation. The algorithm for solving the POMDP, including derivation of the I-space models and discretisation of the continuous I-state POMDP, is described in Section 3.4, while Section 3.5 outlines how plans are executed. Section 3.6 describes a relatively simple environment, BlockWorld, which is used in this and subsequent Chapters to compare algorithms, and explains how Basic-PPOMDP, an MDP-based algorithm, and the PERSEUS algorithm [103] are applied to that world. The results are presented and compared in Section 3.7. PERSEUS was chosen for a comparison because it is a recent algorithm which has shown superior scalability to previous gradient-based methods, and has code available online. Section 3.8 concludes.

3.1 Choosing an Information Space

There are many options for parametric representations, including all the I-spaces deriving from $\mathcal{I}_{prob}$ shown in Figure 2.5. Technically, $\mathcal{I}_{probdisc}$ can be considered a parametric representation of continuous distributions: the set of probability distributions which are piecewise constant over the areas defined by each discrete state. PPOMDP applied to $\mathcal{I}_{probdisc}$ is essentially equivalent to the grid-based methods described in Section 2.5.3. Other parametric representations include the coefficients of series approximations and (possibly mixtures of) any statistical distribution. A good choice of parameterisation is one in which

1. the number of parameters is relatively small; and

2. the class of beliefs likely to occur during plan execution can be well approximated by the chosen parametric form.

The first requirement is important because the dimensionality of the belief space is equal to the number of parameters. Assuming that likely beliefs are uniformly spread over the belief space, the number of beliefs required to achieve a given belief density is exponential in the number of dimensions: the so-called "curse of dimensionality". In practice, of course, the assumption of uniformly likely beliefs is not generally true. This idea is explored further in Chapter 6.

The second requirement is clearly important. The POMDP agent's plans will be useless if they do not consider the types of beliefs which are likely to occur in practice.

These two requirements usually represent competing objectives. Choosing too complex a parametric form, with many parameters, allows likely beliefs to be accurately approximated but will result in onerous computational requirements. Choosing too simple a parametric form will result in a fast planner, but one which produces poor plans. The following section argues that a Gaussian approximation provides a good balance.

3.2   Gaussian Information Spaces

This section argues that Gaussians provide a good approximation for beliefs likely to occur most often during robot navigation, while providing a representation sufficiently compact to make planning tractable.

3.2.1   Quality of a Gaussian Approximation

Beliefs need to be represented and updated for two purposes: firstly for planning, and secondly for online belief tracking (or localisation). This section argues for the use of a Gaussian approximation for planning. It makes the following points, which will be justified below:

1. Gaussians are good approximations for the kinds of beliefs which usually occur during localisation;

2. in realistic long-term localisation problems, the period of time during which an agent holds multimodal beliefs is likely to be relatively short; and

3. isolated instances of multimodal beliefs are far more serious for localisation than for planning.

Under these assumptions, Gaussians are a sensible choice. A disadvantage is their inability to represent multimodal beliefs. We argue that it is critical for a localiser be able to accurately track all beliefs which may occur, even if they are unlikely, because catastrophic localisation failure can occur otherwise. In contrast, non-representable beliefs for a planner may result in sub-optimal behaviour. Occasional sub-optimality can be justified if it results in tractable planning which out-performs simpler heuristic planners.

Gaussians for Localisation

Gaussian distributions have proven to be a good model for probability distributions which occur in practical robot navigation systems [35][7][57]. In a linear world with additive white Gaussian process and observation noise models, the application of a Bayesian belief transition function f_I to a Gaussian prior induces a Gaussian posterior. In nonlinear worlds, an approximation to the Bayesian update, based on linearisation about the mean, produces a Gaussian approximation to the true posterior. This is the basis for Extended Kalman Filter (EKF) based robot navigation algorithms, which have been implemented in many real environments (see for example the survey paper [35]) and have been in use for many years in industrial applications [34].

Occurrence of Multimodal Beliefs

We argue that multimodal beliefs are relatively infrequent in extended robot navigation tasks. Robot navigation tasks are often initialised with a uniform (or at least broad) prior belief. The usual course of events is that the localisation filter undergoes a period of convergence to a unimodal belief, from which the robot proceeds to carry out its task. There are therefore two scenarios in which multimodal beliefs can occur: during or after this global localisation phase.

In any kind of long term robot navigation task, this initialisation phase represents a small fraction of the entire time spent navigating. While the time required to complete the global localisation problem depends on the environment, sensors, and actions, it can certainly be completed in less than five minutes in domains where robust localisation has been shown[1]. For the real-world application domain introduced in Chapter 1, a particle filter requires on the order of a few seconds to collapse to a single mode, as compared with a desired operating duration of up to eight hours. This application will be developed in more detail in Chapter 8.

The second scenario where multimodal beliefs can occur is after global localisation, due to a unimodal belief diverging into separate modes. While the problem is slightly different from localisation, the extensive literature on successful EKF-based SLAM [35] provides evidence that this occurrence is relatively infrequent and short-lived. The SLAM problem begins from a unimodal belief, and the EKF formulation requires that this belief remains unimodal. One potential problem for SLAM is uncertain data association, which could be handled by maintaining multimodal beliefs. The success of batch association methods for EKF-based systems [7] is evidence that uncertain associations (and hence multimodal beliefs) can be resolved quickly in most environments.

The Effects of Multimodal Beliefs

In order to guarantee a unimodal posterior, an EKF-based localiser must associate each observation with exactly one candidate feature. If even a single observation is fused according to an incorrect hypobook, the filter can fail catastrophically [7][50]. Therefore in a reasonably complicated environment, the ability to track multimodal beliefs can be critical. Even if such beliefs are rare in practice, the fact that such a rare event can cause total failure demands that the belief tracker be prepared.

In contrast, an occasional multimodal belief is not catastrophic in the same way for Gaussian-based planning. Rather than total failure, the occurrence of multimodal beliefs may result in behaviour which is occasionally sub-optimal. This can be justified if it results in planning being

---

[1]Two examples from the literature quote the distance of travel required for the completion of global localisation, using sonar sensors indoors, as 2m [44] and 55m [42]. The latter was around a featureless loop.

tractable. A heuristic approach to dealing with multimodal beliefs may be helpful, especially during global localisation. Chapter 9 discusses possible heuristics.

Unimodal Non-Gaussian Beliefs

While we argue that multimodal beliefs are relatively rare, unimodal non-Gaussian beliefs are more common. However, results will be presented to show that Gaussians provide a sufficiently close approximation to allow good policies.

If the environment and sensor suite is such that the ability to represent multimodal beliefs is deemed necessary, more complex functions can be approximated arbitrarily accurately by mixtures of Gaussians [3][58]. The problem with this approach is that the dimensionality of the sufficient statistics increases linearly with the number of Gaussians, and therefore the number of belief samples required to achieve a given density increases exponentially, as described previously. The potential to represent sums of Gaussians will be discussed in Section 9.2.

3.2.2   Dimensionality of the Belief-Space

For a robot whose pose is described by a reasonably low-dimensional vector, the dimensionality of the Gaussian describing its belief distribution will be fixed and relatively low. This is in contrast to algorithms relying on a discretisation of the state-space, which have a relatively high-dimensional representation which scales with the physical size of the environment.

To illustrate, consider a one-dimensional toy POMDP problem. Discretising the space into $|S|$ cells requires the evaluation of a value function in the $|S|$-dimensional continuous space of distributions over those cells. This becomes expensive for large $|S|$. Instead, one could represent the distribution as a Gaussian with parameters (μ, σ), resulting in a problem of computing a value function over a two-dimensional continuous space.

In terms of computation, this reduction in dimensionality comes at the cost of the inability to apply gradient-based solution methods. However, FVI may be more appropriate for robot navigation problems. It will be shown in Section 3.4.1 that FVI-based approaches have certain advantages when faced with continuous high-dimensional observation spaces. Assuming an FVI-based solution method is adopted, the reduction in dimensionality suggests that a Gaussian approximation will be capable of scaling to physically larger state spaces.

3.3   Function Approximation and Belief Set Selection

As mentioned in Section 2.6, a trade-off exists between the computational cost of applying a function approximator, and freedom in being able to choose the makeup of B. This chapter accepts a rigid constraint on B, namely that the beliefs must lie on a fixed-resolution axis-aligned grid. The function approximation scheme can therefore take advantage of the fact that the belief-space can be split into a grid of boxes, with a datapoint on each corner of each box. The constraint of a fixed-resolution grid will be relaxed in Chapter 6.

The function approximator should make two guarantees: firstly that the approximated value at each corner is equal to the value explicitly stored by the datapoint at that corner, and secondly that the interpolated surface is globally continuous. The second guarantee precludes discontinuous jumps at the junctions between boxes, for example. Two approaches to implementing such a function approximator are multilinear interpolation and an interpolation based on a Freudenthal triangulation [74], discussed in Sections 3.3.1 and 3.3.2 respectively. Chapter 6 extends regular grids to arbitrary belief sets.

3.3.1 Multilinear Interpolation

In a d-dimensional space, each box has 2^d corners. A multilinear interpolation estimates the value of each point in the continuous belief-space as a weighted average of these 2^d points. In the one-dimensional case, multilinear interpolation is equivalent to linear interpolation. In a d-dimensional space, a simple algorithm for performing multilinear interpolation is as follows [31]:

1. pick an arbitrary axis;

2. project the query point along that axis to the two opposing faces of the box, producing two new points;

3. perform two $(d-1)$-dimensional multilinear interpolations to find the values of these points, using the $2^{(d-1)}$ points on each face;

4. set the value of the query point by performing a one-dimensional linear interpolation between the interpolated values of those two points.

The problem with multilinear interpolation is that it requires the examination of every one of the 2^d bounding datapoints.

3.3.2 Freudenthal Triangulation

A Freudenthal triangulation allows an interpolation to be performed in $O(d \log d)$ time, examining only $d+1$ of the datapoints, while providing the two guarantees above (namely global continuity and fitting the datapoints exactly). It is based on the division of each box into

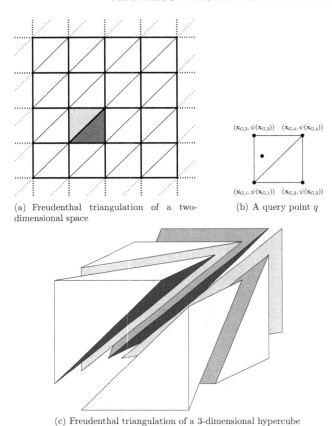

(a) Freudenthal triangulation of a two-dimensional space

(b) A query point q

(c) Freudenthal triangulation of a 3-dimensional hypercube

Figure 3.2: Freudenthal triangulation. (a) shows the Freudenthal triangulation of a two dimensional space. The thick lines show the original hypercubes. Each hypercube contains two hyper-triangles, or simplices, corresponding to the ordering of the two dimensions in paths from the lower-left to upper-right corners. Focussing on the shaded hyper-cube, the upper triangle (the path traverses y then x) contains all points for which $y > x$, while the lower triangle (the path traverses x then y) contains all points for which $x > y$. The value of the two-dimensional query point q shown in (b) can be expressed as a convex combination of the values $\psi(\mathbf{x}_{G,1})$, $\psi(\mathbf{x}_{G,3})$, and $\psi(\mathbf{x}_{G,4})$, stored at $\mathbf{x}_{G,1}$, $\mathbf{x}_{G,3}$, and $\mathbf{x}_{G,4}$. (c) shows how a three-dimensional hypercube is decomposed into $3! = 6$ hyper-triangles (adapted from [75]).

$d!$ hyper-triangles, or *simplices*. Figure 3.2 shows the Freudenthal triangulation of two and three-dimensional spaces.

The triangulation of each box can be performed as follows. First, translate and scale the box such that it is the unit hypercube, with diagonally opposite corners lying on $(x_1, x_2, \ldots, x_d) =$

$(0, 0, \ldots, 0)$ and $(1, 1, \ldots, 1)$. Second, consider all possible paths from $(0, 0, \ldots, 0)$ to $(1, 1, \ldots, 1)$ along the (axis-aligned) edges of the box. There are $d!$ such paths, each consisting of $d+1$ points. The convex hull of each path defines one of the $d!$ hyper-triangles making up the triangulation. Note that each hyper-triangle corresponds to one possible permutation p of $(1, 2, \ldots, d)$, and bounds the set of points satisfying

$$0 \leq x_{p(1)} \leq x_{p(2)} \leq \cdots \leq x_{p(d)} \leq 1 \tag{3.1}$$

In other words, each hyper-triangle is defined by a permutation of the order in which dimensions are traversed in paths between opposing corners, and bounds the set of points whose coordinates obey a particular inequality relationship. Figure 3.2(a) illustrates this with a two-dimensional example. Finally, re-scale and translate the set of hyper-triangles back to their original positions.

It is possible to perform an interpolation using this triangulation without ever explicitly generating all $d!$ simplices [31]. Assuming a query point q defined by the coordinates $(x_1, \ldots, x_d)$, this interpolation can be performed as follows:

1. translate and scale q's bounding box such that it is the unit hypercube, transforming the coordinates of q to $(x'_1, \ldots, x'_d)$;

2. sort the coordinates x'_1 though x'_d from largest to smallest. This identifies the bounding simplex, or hyper-triangle (using Equation 3.1);

3. produce a set of coefficients by expressing $(x'_1, \ldots, x'_d)$ as a convex combination of the coordinates of the bounding simplex's $(d + 1)$ corners; and

4. use the coefficients determined in the previous step as the weights for a weighted sum of the data values stored at the corresponding corners.

For a detailed explanation of the third step, see [75]. The computational cost of this algorithm is dominated by the sorting in the second step, which can be achieved in $O(d \log d)$ time, a significant improvement on multilinear interpolation's cost of $O(2^d)$.

3.4 Solving Parametric POMDPs

This section describes how to derive the I-state MDP models, and how to discretise and solve the MDP. In order to give the high-level view first, it begins with the assumption that the I-state MDP models are already defined. Section 3.4.1 will discuss the derivation of these models, and Section 3.4.2 will analyse the computational complexity of solving PPOMDPs.

Assuming the I-state MDP is defined, two steps are required to solve it:

1. Use FVI to discretise the continuous I-state MDP:

    (a) Calculate the discrete transition function T

    (b) Calculate the discrete reward function R

2. Solve the discrete I-state MDP.

The second step, of solving the discrete I-state MDP, is relatively simple. The POMDP problems considered in this document give rise to discrete I-state MDPs small enough to be represented explicitly, and hence can be solved straightforwardly using Algorithm 1. It will be shown that this represents a small fraction of the total computational cost. The first step, of converting to a discrete MDP, is much more demanding.

The fitted value iteration discussion from 2.5.1 defined the state transition probabilities using Equation 2.55, reproduced here:

$$T(\mathbf{x}_{G,i}, \mathbf{u}, \mathbf{x}_{G,j}^+) = \sum_{\mathbf{w}^+} p(\mathbf{w}^+ | \mathbf{x}_{G,i}, \mathbf{u}) \lambda_G(f(\mathbf{x}_{G,i}, \mathbf{u}, \mathbf{w}^+), j) \qquad (3.2)$$

Applying this to an I-state MDP simply involves replacing $\mathbf{x}$ with $\mathbf{I}$, r with $r_\mathbf{I}$, $\mathbf{w}$ with $\mathbf{z}$, f with $f_\mathbf{I}$, and the state set G with B:

$$T(\mathbf{I}_{B,i}, \mathbf{u}, \mathbf{I}_{B,j}^+) = \sum_{\mathbf{z}^+} p(\mathbf{z}^+ | \mathbf{I}_{B,i}, \mathbf{u}) \lambda_B(f_\mathbf{I}(\mathbf{I}_{B,i}, \mathbf{u}, \mathbf{z}^+), j) \qquad (3.3)$$

where T is an explicit table of transition probabilities representing $p(\mathbf{I}_B^+ | \mathbf{I}_B, \mathbf{u})$, for each I-state in B.

The discrete transition and reward functions can be calculated using Algorithms 2 and 3 respectively. Algorithm 2 can be seen as an agent exercising a mental simulator of the world. For each belief $\mathbf{I}_B$ in B, it resets the simulator to $\mathbf{I}_B$, then considers the actions and observations. For each action-observation pair, it simulates the world forward one step to see which other beliefs in B are (approximately) reachable, then resets the simulator back to $\mathbf{I}_B$.

3.4.1   Deriving the I-State MDP

The implementation of Algorithms 2 and 3 requires the I-state reward, transition and observation functions. This Section describes the derivation of these functions and the computational complexity of Algorithms 2 and 3.

Algorithm 2 Discretising the transition function of a Parametric POMDP. This algorithm outputs the discrete conditional probability table T, representing $p(\mathbf{I}_B^+|\mathbf{I}_B, \mathbf{u})$, for all I-states in B.

---
1   $T(\mathbf{I}_B, \mathbf{u}, \mathbf{I}_B^+) \leftarrow 0, \; \forall \mathbf{I}_B \in B, \; \forall \mathbf{u} \in U, \; \forall \mathbf{I}_B^+ \in B$
2   *foreach* $i \in 1 \ldots |B|$
3   *foreach* $\mathbf{u} \in U$
4   *foreach* $\mathbf{z}^+ \in Z$
5   calculate the probability $p(\mathbf{z}^+|\mathbf{I}_{B,i}, \mathbf{u})$
6   calculate $\mathbf{I}^+ \leftarrow f_{\mathbf{I}}(\mathbf{I}_{B,i}, \mathbf{u}, \mathbf{z}^+)$
7   *foreach* $j \in 1 \ldots |B|$
8                   $T(\mathbf{I}_{B,i}, \mathbf{u}, \mathbf{I}_{B,j}^+) \leftarrow T(\mathbf{I}_{B,i}, \mathbf{u}, \mathbf{I}_{B,j}^+) + p(\mathbf{z}^+|\mathbf{I}_{B,i}, \mathbf{u})\lambda_B(\mathbf{I}^+, j)$
9   *end foreach* j
10   *end foreach* $\mathbf{z}^+$
11   *end foreach* $\mathbf{u}$
12 *end foreach* i

Algorithm 3 Discretising the reward function of a Parametric POMDP. This Algorithm outputs the discrete reward function R for all I-states in B.

---
1   *foreach* $i \in 1 \ldots |B|$
2   *foreach* $\mathbf{u} \in U$
3           $R(\mathbf{I}_{B,i}, \mathbf{u}) \leftarrow R_{\mathbf{I}}(\mathbf{I}_{B,i}, \mathbf{u})$
4   *end foreach* $\mathbf{u}$
5   *end foreach* i

Reward Function

The reward for the I-state $\mathbf{I}$ and action $\mathbf{u}$, used in step 3 of Algorithm 3, can be calculated by an integration over state-space:

$$R_{\mathbf{I}}(\mathbf{I}, \mathbf{u}) = \int_{\mathbf{x}} p(\mathbf{x}|\mathbf{I}) R(\mathbf{x}, \mathbf{u}) d\mathbf{x} \tag{3.4}$$

In the absence of an analytic solution, this Equation can be evaluated using Monte Carlo methods.

Transition Function

The I-state transition function, used in step 6, is in principal based on the Bayesian update given in Equation 2.28. However, the constraint of remaining in the parametric I-space means that some approximation to these Equations is required. Additionally, since step 6 is inside a loop that will be executed many times, ideally it should be possible to calculate $f_{\mathbf{I}}$ efficiently. In this Chapter, we take advantage of the specifics of the example problem to implement an efficient $f_{\mathbf{I}}$, as described in Section 3.6.3. Chapter 4 describes a more general approach.

Observation Function

In the case of large discrete or continuous observation spaces, a complete summation over all possible observations in step 4 may not be possible. One possible solution, as suggested by Roy [92], is to condition on the current and next state:

$$p(\mathbf{z}^+|\mathbf{I}_{B,i}, \mathbf{u}) = \int_{\mathbf{x}^+} p(\mathbf{z}^+|\mathbf{x}^+, \mathbf{u}) \int_{\mathbf{x}} p(\mathbf{x}^+|\mathbf{x}, \mathbf{u}) p(\mathbf{x}|\mathbf{I}_{B,i}) \ d\mathbf{x}d\mathbf{x}^+ \qquad (3.5)$$

Equation 3.5 can be evaluated using Monte Carlo methods, as follows. First, sample a state $\mathbf{x}$ from $p(\mathbf{x}|\mathbf{I}_{B,i})$. Second, given the current state, sample a predicted state $\mathbf{x}^+$ from the process model, $p(\mathbf{x}^+|\mathbf{x}, \mathbf{u})$. Third, given the predicted state, sample from the underlying observation model $p(\mathbf{z}^+|\mathbf{x}^+, \mathbf{u})$.

This approach produces a set of equally-likely observation samples, rather than a complete enumeration of observations with different probabilities. Recognising this, and using N to denote the number of observation samples, Algorithm 2 can be replaced with Algorithm 4, for use with continuous or large discrete observation spaces.

Algorithm 4 This Algorithm represents a modification of Algorithm 2, for use with continuous or large discrete observation spaces. N denotes the number of observation samples.

---
1    $T(\mathbf{I}_B, \mathbf{u}, \mathbf{I}_B^+) \leftarrow 0, \ \forall \mathbf{I}_B \in B, \ \forall \mathbf{u} \in U, \ \forall \mathbf{I}_B^+ \in B$
2   *foreach* $i \in 1 \ldots |B|$
3   *foreach* $\mathbf{u} \in U$
4   *foreach* $n \in 1 \ldots N$
5   sample a state $\mathbf{x}$ from $p(\mathbf{x}|\mathbf{I}_{B,i})$
6   sample a predicted state $\mathbf{x}^+$ from $p(\mathbf{x}^+|\mathbf{x}, \mathbf{u})$
7   sample an observation $\mathbf{z}^+$ from $p(\mathbf{z}^+|\mathbf{x}^+, \mathbf{u})$
8   calculate $\mathbf{I}^+ \leftarrow f_{\mathbf{I}}(\mathbf{I}_{B,i}, \mathbf{u}, \mathbf{z}^+)$
9   *foreach* $j \in 1 \ldots |B|$
10                    $T(\mathbf{I}_{B,i}, \mathbf{u}, \mathbf{I}_{B,j}^+) \leftarrow T(\mathbf{I}_{B,i}, \mathbf{u}, \mathbf{I}_{B,j}^+) + \frac{1}{N}\lambda_B(\mathbf{I}^+, j)$
11   *end foreach* j
12   *end foreach* n
13   *end foreach* $\mathbf{u}$
14   *end foreach* i

Conditioning on the set of likely states in this way has the effect of focussing computation on the set of likely observations given the belief. This is in contrast to gradient-based methods, which must define a set of per-state observation probabilities for a fixed global set of observations ($p(\mathbf{z}^+|s^+, \mathbf{u})$ in Equation 2.37). Figure 2.6 makes the distinction clear: the first step of the gradient-based approach uses the world model to generate α-vectors, while particular beliefs are not exposed until a second step which selects α-vectors. This represents both a strength and a weakness of gradient-based approaches. The linearity of the value function means that α-vectors generalise over the belief-space, and hence FVI's iteration over beliefs is not required.

However, in order to generalise over the entire belief-space, conditioning on particular beliefs (and hence focussing on likely observations) is not possible.

3.4.2 Computational Complexity

The following discussion assumes that Algorithm 4 is used, hence the number of iterations over observation-space is given by N. For a discrete observation-space, N can be replaced by $|Z|$.

When analysing complexity, note that steps 9-11 of Algorithm 4 needn't iterate over every belief in B. Rather, they need iterate over only those for which $\lambda_B(\mathbf{I}^+, j)$ is non-zero. In other words, steps 9-11 need examine only the set of beliefs reachable from $\mathbf{I}_{B,i}$. If T is sparse, this set will be much smaller than B. Therefore, let $C(\lambda_B)$ denote the cost of calculating the weighting function, and let $|\lambda_B|$ denote the average number of non-zero weightings returned. Letting $C(f_\mathbf{I})$ denote the complexity of the belief transition function and $C(R_\mathbf{I})$ denote the complexity of the I-state reward function, the total computational complexity of Algorithm 2 is

$$O\big(|B||U|N(C(f_\mathbf{I}) + C(\lambda_B) + |\lambda_B|)\big) + O(|B||U|C(R_\mathbf{I})) \tag{3.6}$$

Unless the reward function is particularly expensive, the complexity of discretising an I-state MDP will be dominated by the first term, which generates T. If, in addition, it is assumed that the cost of calculating the transition function is significantly larger than the cost of calculating weights or updating T, this cost can be approximated by

$$O\big(|B||U|NC(f_\mathbf{I})\big) \tag{3.7}$$

Note that this complexity is only dependent on the size of the state-space through $|B|$, the number of belief points required to cover that state-space.

In comparison, an update of a discrete gradient-based method with a point-based approximation has complexity

$$O(|\Gamma^+||S|^2|U||Z| + |\Gamma^+||S||U||Z||B|) \tag{3.8}$$

(see Section 2.4.2). The dependence on $|S|^2$ represents the cost of applying the transition matrix T, and can often be closer to $|S|$ with the use of sparse matrix methods. Regardless, the important thing to note is that algorithms such as PBVI [86] and PERSEUS [103] scale with the size of the state-space in addition to scaling with the number of beliefs required to cover that state-space. Section 3.7.2 will demonstrate the effect of this experimentally by comparing PPOMDP planning against PERSEUS on progressively larger state-spaces.

3.5   Plan Execution

Plan execution requires two components: belief tracking and action selection. While a belief transition function $f_{\mathbf{I}}$ is required to perform belief updates during planning, there is no reason in principle why online belief tracking should be performed with the same function. In fact, there may be good reasons to use a different algorithm. $f_{\mathbf{I}}$ operates entirely in a particular I-space, which was chosen for the reasons outlined in Section 3.2. Considerations such as the number of parameters are important for planning, but of minor importance for online belief tracking. A better approach may be to track beliefs online in a more complicated I-space (such as the space of sums of Gaussians), mapping to the closest belief in the planning I-space whenever a decision is required.

Even when the planning and online belief-tracking I-spaces are identical, the efficiency concerns of the online tracker are different from the planner. It may be appropriate to make different approximations in the two scenarios. For the BlockWorld problem which will be presented in Section 3.6, however, the same belief transition function is used for both planning and belief tracking. In Chapter 8, when executing plans in a real environment, a more sophisticated belief tracker is used.

Actions are selected during plan execution based on the value function. The value function gives the expected discounted cumulative reward, over an infinite horizon, for every possible belief. Armed with a value function, an agent need not plan ahead when encountering a belief online, since that value function implicitly encodes the results of prior planning. Instead, it is sufficient to simply 'surf' the value function with a one-step lookahead. That is, it is sufficient from belief $\mathbf{I}$ to apply the policy

$$\pi_{\mathbf{I}}(\mathbf{I}) = \arg\max_{\mathbf{u}} R_{\mathbf{I}}(\mathbf{I}, \mathbf{u}) + \gamma \mathop{E}_{\mathbf{z}^+}\big[V(f_{\mathbf{I}}(\mathbf{I}, \mathbf{u}, \mathbf{z}^+)) \big] \qquad (3.9)$$

Equation 3.9 entails some online computational cost, since the agent must calculate the belief transition function $|U||Z|$ times in order to choose an action. A cheaper alternative is to perform zero-step lookahead. Since value iteration requires a maximisation over actions (see Algorithm 1), one can store the maximising action for each belief and simply apply it online during plan execution. The only complication arises from the fact that the belief-space is continuous. In the context of PPOMDP, value iteration produces maximising actions for every belief in B, but $\mathbf{I}$ may lie anywhere in the continuous belief-space. A simple approach is to select the maximising action for the belief in B which is nearest to the current belief (as determined by λ_B). Unless otherwise stated, this is the approach taken for the PPOMDP algorithm throughout this document. The quality of control using this zero-step lookahead, as compared to one-step lookahead, or even n-step lookahead as suggested in Section 2.7.3, will be discussed in detail in Chapter 7.

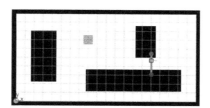

Figure 3.3: The basic continuous navigation environment: a 20m×10m hall, with obstacles shown in black. The goal region is indicated by the 1m×1m dark cyan square in the central open area. The robot, to the right of the hall, has four sensors which noisily detect the ranges to obstacles in each of the four compass directions. The noisy outputs of the north and south sensor ranges are shown. The 1m×1m grid shows the state space discretisation used to evaluate the MDP and PERSEUS algorithms.

3.6 BlockWorld: A Simple Continuous Navigation Problem

Since many of the benchmark POMDP problems from the literature assume a discrete state space, comparison against the state of the art is difficult. Rather than invent an entirely new problem, a comparison is performed by modifying the continuous navigation problem to which the PERSEUS algorithm was applied in [103]. Section 3.6.1 describes the rules of BlockWorld, then Sections 3.6.2 and 3.6.3 outline the application of three planning algorithms to those rules: PPOMDP, PERSEUS, and an MDP-based heuristic. The performance of all three algorithms is evaluated using a simulator which implements the continuous version of the world defined in Section 3.6.1.

3.6.1 The Rules of BlockWorld

The continuous state space is the 20m×10m hallway shown in Figure 3.3. The action space of the simulated robot is the continuous set of (d, θ) pairs, where the travel distance d and heading θ parameters are in the ranges $[0,2]$ metres and $(-\pi, \pi]$ radians respectively. In the absence of obstacles, the transition function $p(\mathbf{x}^+|\mathbf{x}, \mathbf{u})$ is a Gaussian distribution whose mean is determined by translating the previous pose d metres in the direction θ, and whose standard deviation is given by $0.2 \times d\mathbf{I}$, where $\mathbf{I}$ in this case refers to the identity matrix. If the path from the previous pose to the next pose would collide with an obstacle, the robot remains stationary. For the purposes of calculating collisions, the robot is considered to be a point mass.

The robot is equipped with four range sensors, with one pointed in each of the four absolute compass directions. Each sensor will detect an obstacle if it is within the sensor's maximum range of 2m. The range reported by the sensor is determined by a Gaussian distribution, centred on the true range, with variance 0.5m. In addition, the robot has a collision sensor which indicates whether or not the previous action was blocked by an obstacle.

The robot's performance is evaluated in a series of episodes. Each episode begins from a random valid state. The robot is given an initial belief with a variance in both the x and y position estimate of 1m. The mean of this initial belief is sampled from a Gaussian distribution, with a variance in x and y of 1m, centred on the initial state. The episode ends after 100 actions have been taken or after the goal has been reached, whichever occurs first. The goal region is the 1m×1m square area shown in Figure 3.3. The robot receives a reward of $+10$ for an action which brings it to the goal, and -0.1 for any other action. The reward attainable over an episode is therefore in the range $[-10, 10]$. All algorithms were evaluated using the same continuous world simulator.

For all solution algorithms the action space is discretised. 33 actions are allowed: the agent can choose from 16 headings spaced equally over the range $(-\pi, \pi]$, combined with a distance of either one or two metres. The 33rd action is $(0.1, 0)$, allowing the robot to make small (noisy) motions in the vicinity of the goal. The fineness of this discretisation should allow an effective planner to generate a good policy.

3.6.2 Discrete Solutions

Of the three solution algorithms (MDP, PERSEUS, and PPOMDP), this section describes MDP and PERSEUS. To apply these solution algorithms, the state space was first partitioned with a regular two-dimensional grid as in Figure 3.3. When discretising the observation space, the number of observations is equal to $2 \times 4^{n+1}$ where the leading 2 is the number of distinct outputs from the (binary) collision sensor, 4 is the number of range sensors, n is the number of bins into which the range of each sensor is discretised, and the $+1$ in the exponent accounts for the fact that the range sensors may sense no range. All experiments used $n = 1$ (giving 32 discrete observations), which essentially meant ignoring the range information. A finer discretisation would improve results, but at the cost of an exponential increase in the number of observations and hence in running time. The validity of ignoring the range information will be discussed shortly, in comparison with the parametric solution.

The discrete versions of the transition and observation functions, namely $p(s^+|s, \mathbf{u})$ and $p(\mathbf{z}^+|s^+, \mathbf{u})$ were determined by sampling. For each discrete cell s and action $\mathbf{u}$, 50 continuous state samples were drawn from a uniform distribution over the area of that cell. For each continuous state sample, 5 samples were drawn from the distribution $p(\mathbf{x}^+|\mathbf{x}, \mathbf{u})$. This procedure gives a continuous distribution $p(\mathbf{x}^+|s, \mathbf{u})$ which was then mapped back into discrete space, by counting the number of samples within each cell, to give $p(s^+|s, \mathbf{u})$. $p(\mathbf{z}^+|s^+, \mathbf{u})$ was evaluated similarly.

The reward for a state, $R(s, \mathbf{u})$, is action-independent and was determined by integrating the continuous reward function over the discrete state's area.

Discrete MDP Solution

The MDP version of the discretised problem is relatively simple to solve. The state is assumed to be fully observable and equal to the maximum-likelihood state (this is technically the *MLS* heuristic, as presented in Section 2.7.1). The observation probabilities are therefore ignored for planning, but are still useful for tracking the discrete belief-state during plan execution. The MDP policy π^* can be found from the transition and reward matrices, $p(s^+|s, \mathbf{u})$ and $R(s, \mathbf{u})$, by value iteration performed using Algorithm 1. The running time of the algorithm was measured by adding the CPU time spent calculating the transition matrices to the time spent during value iteration.

PERSEUS Solution

While the PERSEUS algorithm is described briefly in Section 2.4.2, readers are directed to [103] for the detailed mechanics of the algorithm. Given the discrete transition, observation, and reward matrices described earlier in this section, the only remaining free parameter is the number of belief points and the algorithm by which they are selected. 10000 belief points were chosen by forward simulation of the model using random actions from random initial conditions.

Again, running times were measured as the sum of the time spent calculating transition matrices and the time spent during value iteration. The former was constant and relatively small, taking 3.7 seconds for a 20×10 discretisation of the state space. Note that the time spent establishing a belief set by forward simulation of the model is not included. This is in contrast to subsequent chapters, in which the time spent establishing a belief set for PPOMDP is included in overall running times.

3.6.3 Parametric Solution

Basic-PPOMDP represents beliefs as two-dimensional Gaussians with diagonal covariance matrices, giving rise to a four-dimensional belief space (two for the mean, two for the diagonal covariance matrix). Belief points were chosen on a regular grid. The means of the belief points were chosen on a 21×11 grid, while the diagonal elements of the covariance matrix were discretised into six levels from 0.1 to 4.0 inclusive, giving a total of 8316 belief points. Linear interpolation using a Freudenthal triangulation was chosen as a function approximation scheme, as described in Section 3.3.

The belief transition function $f_I(\mathbf{I}, \mathbf{u}, \mathbf{z}^+)$ was relatively simple. If the collision sensor does not register a collision, the mean is shifted and the covariance expanded according to the transition function described earlier in this section. When introducing range information from one of the four range sensors, the problem is treated as two de-coupled one-dimensional estimation problems. For each sensor, a reading can only have come from one of four known features in the world: the edge of one of the three obstacles or the far wall. After solving a relatively simple data association problem, the range can be viewed as a direct observation of the robot's position in one dimension. The robot's mean μ_R and uncertainty σ_R^2 in the dimension along which the sensor can sense is updated using

$$\mu_{R'} = (\sigma_R^2 + \sigma_S^2)^{-1}(\mu_R \sigma_S^2 + \mu_S \sigma_R^2) \tag{3.10}$$

$$\sigma_{R'}^2 = (\sigma_R^2 + \sigma_S^2)^{-1}(\sigma_R^2 \sigma_S^2) \tag{3.11}$$

where μ_S is the robot's position as estimated by the sensor and σ_S^2 is the sensor's variance.

This sensor update scheme clearly ignores some information, since an observation of one of the central obstacles with the north-pointing sensor should constrain the distribution over east-west poses, and a missing observation should confer some information. However it is not entirely clear how to incorporate this information while preserving the Gaussian form and the efficiency of the update, and the update as described is a reasonable approximation most of the time. Section 3.7 shows an example where the sub-optimality of the belief update function results in the robot failing to reach the goal.

In order to calculate T during planning, Equation 3.5 was evaluated using sampling, with $N = 50$ samples. The sampling scheme may seem sparse, however the number of samples is directly related to the computation time and an increase was found, empirically, to have little effect on the quality of plans. It may seem unfair to take the actual values of the ranges into account for the PPOMDP algorithm but not for the discrete versions. However, as pointed out in Section 3.4.1, this is due to a fundamental difference between the algorithms: the discrete versions must calculate the effects of observations without reference to any specific belief point, and therefore fix a set of globally representative observations. Accounting for different ranges would cause an exponential expansion in the size of this set, and therefore in computational requirements. In contrast, the PPOMDP algorithm can choose a different set of representative observations for every belief point considered.

Finally, the reward function, $R_I(\mathbf{I}, \mathbf{u})$, is independent of the action. It is

$$R_I(\mathbf{I}, \mathbf{u}) = r_g G + r_{\bar{g}}(1 - G) \tag{3.12}$$

where $r_g = 10$ is the reward for reaching the goal, $r_{\bar{g}} = -0.1$ is the reward associated with any

	PPOMDP	PERSEUS	MDP
Number of Discrete States	n/a	200	200
Number of Discrete Beliefs	8316	10000	n/a
Number of Discrete Actions	33	33	33

Table 3.1: A summary of the parameters used for the three algorithms.

other state, and G is the volume integral of the belief distribution over the goal, evaluated by sampling.

Of the steps involved in generating the Basic-PPOMDP solution, the computational cost of evaluating Algorithm 4, involving repeated application of the belief transition function, dominates. This highlights the need for an efficient belief transition function. Actual value iteration consumed only 3% of the total running time, with discretisation of the MDP requiring the majority. This is in contrast to the PERSEUS algorithm, whose running time is dominated by value iteration.

3.6.4   Parameter Summary

The parameter settings are summarised in Table 3.1. 200 states and 10000 beliefs were chosen for PERSEUS because these numbers were used for the original problem from which BlockWorld is derived [103]. While the original problem selected states by clustering training data, states were selected on a regular grid in this problem for simplicity. PPOMDP's discretisation of the belief-space in x and y was chosen to match the discretisation of the state-space used for PERSEUS. The discretisation of the variances was chosen to give PPOMDP and PERSEUS a similar number of beliefs.

The discretisation of the observation space cannot be compared directly. As explained in Section 3.4.1, PERSEUS's discretisation of the observation-space must be fixed before running the algorithm, whereas PPOMDP can sample observations based on each belief.

3.7   Basic BlockWorld Experiments

The algorithms were compared in two scenarios. In the first, the standard BlockWorld problem from Section 3.6.1 was solved. In the second the environment from Figure 3.3 was expanded to test scalability with the size of the state space. All results in this book were produced on the same 2.0GHz Pentium M laptop with 1Gb of RAM.

Experimentation showed that the results are particularly sensitive to the precise location of the goal. If the discrete states (or the means of discrete belief points) happen to be aligned

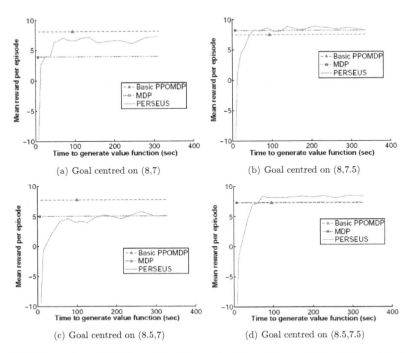

(a) Goal centred on (8,7)   (b) Goal centred on (8,7.5)

(c) Goal centred on (8.5,7)   (d) Goal centred on (8.5,7.5)

Figure 3.4: Comparison of the three algorithms using the environment from Figure 3.3, showing the mean total reward per episode versus the time required to calculate a policy. Each algorithm is tested under the four slightly different goal configurations. Since PERSEUS's running time is dominated by value iteration and a valid plan is available after each iteration, the mean reward attained by every plan is shown. Since the running time of both MDP and PPOMDP is dominated by the pre-computation stage, only a single datapoint, indicated by the marker, is shown for each. The horizontal lines through these markers are simply to facilitate comparison. Each datapoint is the average of 10000 episodes.

with the goal region, the problem is significantly simplified. All comparisons were therefore performed under four conditions: with the centre of the goal located at $(8, 7)$, $(8.5, 7)$, $(8, 7.5)$ and $(8.5, 7.5)$. The first location is perfectly aligned with the discrete state-space, the last is perfectly mis-aligned, while the other two are aligned in one dimension.

3.7.1 Comparison in the Standard World

Results on the standard world are shown in Figures 3.4 and 3.5. For PERSEUS, the figures show how the policy improves over time. This is done by remembering the intermediate value functions generated at each stage of value iteration, and evaluating a policy based on each one.

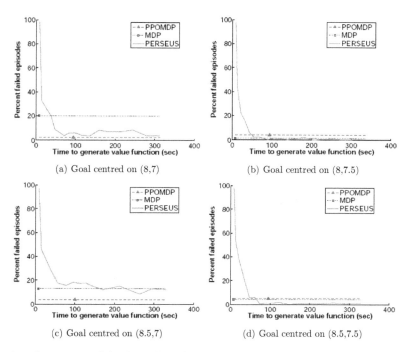

Figure 3.5: Comparison of the three algorithms using the environment from Figure 3.3, showing the percentage of episodes in which the robot failed to reach the goal, versus the time required to calculate a policy. The four plots show the four different goal configurations. As for Figure 3.4, results for all of PERSEUS's intermediate plans are shown. Each datapoint is the average of 10000 episodes.

This is possible because PERSEUS's running time is dominated by the value iteration process. In contrast, value iteration represents a small component of the running times of MDP and PPOMDP. For these algorithms, the quality of plans undergoes a single step change: no plan is available before value iteration commences, and the time between the beginning of value iteration and its convergence is minimal. Therefore Figures 3.4 and 3.5 show only a single datapoint for MDP and PPOMDP.

Figure 3.4 shows the CPU time required for each algorithm to calculate a value function versus the mean reward attained using the policy based on that value function. The figure shows that both PPOMDP and PERSEUS are capable of producing reasonable plans, and that they are able to out-perform MDP by considering and planning for the uncertainty in their state estimate. The time required for PERSEUS to produce a good plan is approximately the same as the time required for Basic-PPOMDP to generate its plan. It should be noted however that

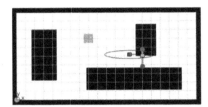

Figure 3.6: A case in which Basic-PPOMDP fails. The belief is indicated by the covariance ellipse with the blue robot at its centre. The true state is indicated by the cyan robot. If the action is blocked by an obstacle, the belief will be updated only along the y-axis.

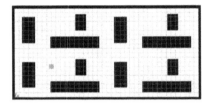

Figure 3.7: An expanded version of the world from Figure 3.3, of size 40m×20m, produced by tiling 2×2 copies of the original world.

the algorithms' running times are subject to the details of their implementations.

Closer examination of the results showed that the agent tends to either reach the goal fairly quickly or become trapped in a state from which it cannot escape for the entire episode. Figure 3.5 shows, for each algorithm, the percentage of episodes in which the robot was unable to reach the goal.

The major cause of becoming trapped for the Basic-PPOMDP agent is the sub-optimality of the belief transition function. Figure 3.6 shows an example. Given the pictured belief, the agent tries to move north-west towards the goal. However since the true state is in the tail of the belief distribution, behind an obstacle, the action is almost certain to fail. Since neither the range sensor nor the collision sensor causes the agent to update its belief along the x-axis, the agent will continue to try exactly the same action.

3.7.2 Comparison in a Tiled World

To evaluate scalability with respect to the size of the state space, the environment was enlarged by "tiling" the world as illustrated in Figure 3.7. Due to memory restrictions, the action-space was reduced to only nine actions. Eight were at 2m, spread uniformly over heading-space, while the ninth was $(0.1, 0)$. Only one goal configuration was tested, centred on $(8, 7)$. While more actions and goals would provide better estimates of the rewards, the aim of the experiment was to test scalability. When using an $n \times m$ tiling the number of belief points per tile was held constant, giving a total of $10000mn$ for PERSEUS and $8316mn$ for Basic-PPOMDP. For the discrete algorithms, the size of a discrete cell remained constant. The number of discrete cells was therefore $200mn$.

The results are shown in Figure 3.8. As the number of tiles is increased, the mean reward of all algorithms decreases. This is expected since the world is physically larger and it therefore takes longer to reach the goal, even for an optimal plan.

The time taken for PPOMDP to generate a plan is both small and linear in the number of belief points (and therefore the number of tiles). While PERSEUS can eventually generate a superior plan for larger worlds, the time required to do so increases rapidly with the number of tiles. While the planning time can be reduced by using a coarser discretisation of the state space, this is likely to result in decreased rewards.

3.8 Summary

This chapter introduced the Basic-PPOMDP algorithm. Section 3.2 argued for planning in the space of Gaussian approximations to arbitrary continuous beliefs, on the basis that they are a good approximation for beliefs which are usually encountered during plan execution, while requiring a relatively small number of sufficient statistics.

The simple weighting function introduced in Section 3.3 uses a Freudenthal triangulation. This allows the examination of far fewer beliefs than a scheme such as multilinear interpolation, however both schemes require that the set of beliefs B lie on a regular grid over belief-space.

Section 3.4 described how to derive the continuous I-state MDP model, and how to use the belief set and weighting function to discretise that model. The algorithm for solving that discrete model was given in Chapter 2. Section 3.5 showed how an agent can select actions when given this solution.

BlockWorld was introduced in Section 3.6. This is a simulated navigation problem on which various algorithms can be compared. It was shown how Basic-PPOMDP and two other algorithms, MDP (which ignores uncertainty) and PERSEUS (a discrete POMDP solution algorithm) can be applied to BlockWorld. A comparison showed that Basic-PPOMDP produced good results when compared to MDP, and reasonable results when compared to PERSEUS. The size of the world was then increased, showing how the planning time required by Basic-PPOMDP scales linearly with the physical size of the world, unlike the time required by PERSEUS.

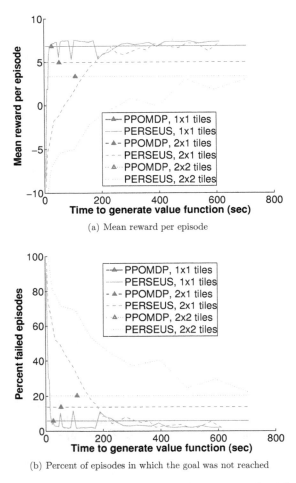

(a) Mean reward per episode

(b) Percent of episodes in which the goal was not reached

Figure 3.8: Comparison of the PPOMDP and PERSEUS algorithms for different sized worlds. For clarity, results are shown for only one goal configuration. Each datapoint is the average of 10000 episodes. Note that the results for a 1×1 tiling do not match previous results due to the different discretisation of the action space.

The following chapters make incremental improvements to Basic-PPOMDP, with each improvement being validated using BlockWorld. The first such improvement, discussed in Chapter 4, addresses problems with Basic-PPOMDP's belief transition function, as highlighted in Section 3.7.1.

Chapter 4

An Improved Belief Transition Function

While the Basic-PPOMDP algorithm performed adequately on the versions of BlockWorld presented in Chapter 3, it has serious deficiencies. These deficiencies are highlighted in Section 4.1, by experimentally applying Basic-PPOMDP to progressively more challenging versions of BlockWorld, showing how performance degrades. The reasons for this degradation are identified in Section 4.2.

Section 4.3 proposes a solution for this problem. It begins by presenting a slightly different view of the PPOMDP algorithm presented in the previous chapter. For each belief point and action, the algorithm presented in the previous chapter can be seen as generating a distribution over posteriors, then approximating that distribution with a set of discrete I-state transitions. Section 4.3.1 presents an improved approach to generating those posteriors, based on particle filtering. While this approach is likely to be more accurate, certain issues are introduced. Firstly, it introduces certain stochastic effects which will be discussed in Section 4.3.2. Secondly, the new belief transition function is not particularly efficient. Section 4.4 presents two approaches to improving its efficiency, based on re-using predictions and re-using likelihood calculations.

Section 4.6 performs an experimental comparison of the improved algorithm with the algorithms evaluated in the previous chapter (MDP, PERSEUS, and Basic-PPOMDP). It shows that planning accuracy is improved significantly over Basic-PPOMDP, while the time required for planning is approximately the same. The computational requirements and scalability of the algorithm are analysed in Section 4.5, and a comparison is made with FVI-based algorithms which rely on an underlying discrete representation. It is shown that, in contrast to algorithms which rely on an underlying discretisation, the cost of the belief transition function proposed in this chapter is independent from the size of the state-space, and hence the algorithm is applicable to large, realistic planning problems. Section 4.7 concludes.

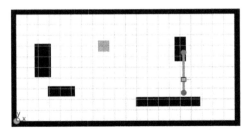

Figure 4.1: A version of the original BlockWorld from Figure 3.3, but with obstacles contracted to reduce obstacle density, thereby making navigation more difficult.

4.1 Experiments on Different Worlds

This section demonstrates how Basic-PPOMDP's performance degrades in more challenging environments. Experiments were carried out in four similar worlds:

1. the original $20 \times 10m$ BlockWorld from Chapter 3;

2. a modified $20 \times 10m$ BlockWorld with smaller obstacles, as shown in Figure 4.1;

3. a larger version of the original BlockWorld, created by scaling Figure 3.3 to $30 \times 15m$; and

4. a $30 \times 15m$ version of the world in Figure 4.1.

Intuitively, the four worlds, in the order listed, should increase in navigational difficulty. Both increasing the size of the world and decreasing the density of obstacles create larger open spaces, making it more difficult for the agent to reach the goal with any certainty.

As in Chapter 3, each world was tested with four goal configurations separated by $0.5m$. The lower-left goal configuration was set to $(8, 7)$ for the smaller worlds and $(10, 9)$ for the larger worlds. For a world of size $x \times y$, MDP and PERSEUS partitioned the state-space into an $x \times y$ grid, while variants of the PPOMDP algorithm used belief points with means on an $x + 1 \times y + 1$ grid (in the centres of the cells used by the discrete algorithms).

Results

The results of comparing Basic-PPOMDP against MDP are shown in Figure 4.2. While Basic-PPOMDP performs reasonably on smaller worlds, its performance clearly degrades as the environment becomes more challenging. The most common cause of failure on larger worlds is that the Basic-PPOMDP agent becomes trapped in cycles. From a well-localised position near

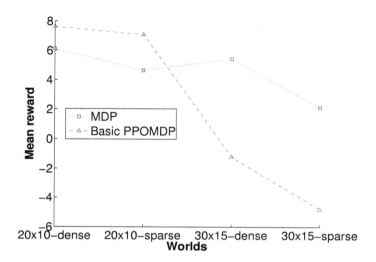

Figure 4.2: Performance of the MDP and Basic-PPOMDP algorithms on four worlds of approximately increasing navigational difficulty. Each datapoint is the mean of four trials (using the four goal configurations), where each trial is the mean of 1000 episodes.

an obstacle, the agent moves into a more open space near the goal. Out of sensor range of the obstacle, its uncertainty grows. Before reaching the vicinity of the goal, however, it turns back, returning to the obstacle in order to improve its localisation. Cycling in this way for the entire episode, it never reaches the goal. The following section will show how this behaviour results from Basic-PPOMDP's inability to accurately predict the likelihood of future observations.

4.2 The Requirement to Anticipate Future Observations

Basic-PPOMDP's parametric belief transition function, presented in Section 3.6.3, is fast but unable to anticipate future observations with sufficient accuracy. This has two obvious manifestations: when the agent becomes trapped in a cycle near an obstacle, as described in the previous section, and when the agent becomes trapped behind an obstacle, as shown in Figure 3.6 and reproduced in Figure 4.3.

The problem when trapped behind an obstacle is the clearer of the two. From the belief shown in Figure 4.3, a move towards the north-west is appropriate. Based on the belief, the chances of a collision are low. However, due to the fact that the true state is in the tail of the distribution, the move is in fact almost certain to fail. The agent will therefore receive a positive observation from the collision sensor, plus range observations from the north and south range sensors.

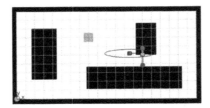

Figure 4.3: A case in which Basic-PPOMDP fails. The belief is indicated by the covariance ellipse with the blue robot at its centre. The true state is indicated by the cyan robot. If the action is blocked by an obstacle, the belief will be updated only along the y-axis.

Unfortunately, the sub-optimal belief transition function cannot incorporate this information, and will update the belief only along the y-axis. In other words, the belief will be approximately the same for the next iteration.

Since the agent predicts the likelihood of future observations purely based on its belief, it has no way of knowing that the same thing is likely to happen on the next iteration. The agent will therefore continue to take the same action, and the situation will persist forever. The only way for the agent to realise that the previous collision implies an increased likelihood of future collisions is if the belief were updated to reflect the previous collision.

The cause for the agent becoming trapped in cycles is its inability to utilise negative information. In other words, it becomes trapped because it is unable to incorporate the information conferred by an observation which was considered possible but did not occur.

Figure 4.4 illustrates an example. After becoming well localised at position a, the agent moves towards the goal. Suppose its true path is directly from a to b. As the agent moves outside sensor-range of the western wall, it will be unable to observe any obstacles. In the absence of observations, the agent's uncertainty will grow to a large, approximately circular ellipse centred on b. Unfortunately this ellipse is a poor approximation to the belief which would result from applying the full Bayesian belief update, given in Equation 2.28. To see why the approximation is poor, consider the shaded probability mass in Figure 4.4, which is within sensor-range of the western wall. This probability mass is invalid. If the true pose were within that area, the western wall would have been observed. Since the wall was not observed, the agent cannot be in that area.

The result of this poor approximation is that the agent cannot accurately predict the probabilities of future observations. As it continues to move towards b in Figure 4.4, it over-estimates the probability of future observations of the western wall. If one were to anthropomorphise the agent, one could say that by the time it reaches b, it considers itself particularly unlucky to have moved so far but not observed the wall. It therefore decides that the best strategy is to

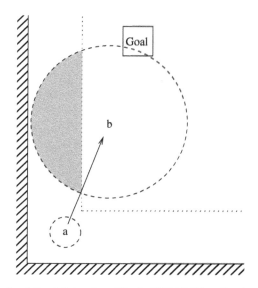

Figure 4.4: An example of the deficiencies of Basic-PPOMDP's estimator. Uncertainty ellipses are represented with dashed lines, and the border of the region from which the walls can be sensed is indicated with a dotted line. Basic-PPOMDP considers the shaded probability mass to be feasible in the absence of observations of the wall to the west, and hence over-estimates the probability of such observations.

return to a in order to re-localise, before trying again in the hope that next time it will not be so unlucky. Of course in reality it was not unlucky, it just has a bad model.

The core of the problem is that the belief transition function picks and chooses which pieces of information to apply, and which to ignore. Both observations of obstacles and non-observations of obstacles potentially confer information, however the belief update function incorporates only the former. More precisely, the strategy is to apply only those pieces of information which result in a Gaussian posterior. While this is a common tactic in EKF-based localisation or SLAM systems [35], the requirements for planning are more stringent: in addition to being able to track its belief, a planning agent must be able to accurately estimate the likelihood of future observations. Clearly then, PPOMDP requires a belief transition function which is efficient, uses all available information, and produces a Gaussian posterior. The following section attempts to provide this.

4.3   An Improved Belief Transition Function

For each belief and action, Algorithm 4 generates a set of discrete transition probabilities. This can be broken into two distinct steps: a first which generates a distribution over posteriors, and a second which uses the weighting function to map that distribution over posteriors to a set of discrete transitions. Let Δ denote a set of equally likely posterior beliefs, and let Δ_l to denote the l'th possible posterior. Algorithm 4 can then be replaced by Algorithms 5 and 6, where Algorithm 5 calls Algorithm 6 to generate each Δ, then translates that Δ into a set of discrete transitions.

Algorithm 5 A different view of Algorithm 4. For each belief and action, first generate a distribution over posteriors Δ, then translate those posteriors into discrete transition probabilities.

---

1   $T(\mathbf{I}_B, \mathbf{u}, \mathbf{I}_B^+) \leftarrow 0, \ \forall \mathbf{I}_B \in B, \ \forall \mathbf{u} \in U, \ \forall \mathbf{I}_B^+ \in B$
2   *foreach* $i \in 1 \ldots |B|$
3   *foreach* $\mathbf{u} \in U$
4           $\Delta \leftarrow$ **generateDistributionOverPosteriors**$(\mathbf{I}_{B,i}, \mathbf{u})$
5   *foreach* $l \in 1 \ldots |\Delta|$
6   *foreach* $j \in 1 \ldots |B|$
7                   $T(\mathbf{I}_{B,i}, \mathbf{u}, \mathbf{I}_{B,j}^+) \leftarrow T(\mathbf{I}_{B,i}, \mathbf{u}, \mathbf{I}_{B,j}^+) + \frac{1}{|\Delta|} \lambda_B(\Delta_l, j)$
8   *end foreach* j
9   *end foreach* l
10   *end foreach* $\mathbf{u}$
11 *end foreach* i

Algorithm 6 The version of **generateDistributionOverPosteriors**$(\mathbf{I}, \mathbf{u})$ used in Section 3.4.1. N is a free parameter specifying the number of observation samples, and hence the number of posteriors.

---

1   $\Delta \leftarrow \emptyset$
2   *foreach* $n \in 1 \ldots N$
3   sample a state $\mathbf{x}$ from $p(\mathbf{x}|\mathbf{I})$
4   sample a predicted state $\mathbf{x}^+$ from $p(\mathbf{x}^+|\mathbf{x}, \mathbf{u})$
5   sample an observation $\mathbf{z}^+$ from $p(\mathbf{z}^+|\mathbf{x}^+, \mathbf{u})$
6   add the posterior $\mathbf{I}^+ = f_\mathbf{I}(\mathbf{I}, \mathbf{u}, \mathbf{z}^+)$ to Δ
7   *end foreach* n

4.3.1   A Belief Transition Function Using Monte Carlo Methods

The problems identified in Section 4.2 stem from the fact that the belief transition function $f_\mathbf{I}$, in step 6 of Algorithm 6, ignores certain types of information. The alternative proposed in this section is to approximate $f_\mathbf{I}$ using sequential Monte Carlo methods [2][4], which have become popular for mobile robot localisation [110]. A distinct advantage of Monte Carlo Localisation

(MCL) is its ability to handle arbitrary process and observation models, including the use of negative information and information which produces non-Gaussian posteriors.

MCL involves using sampling to approximate the belief update given in Equations 2.26 and 2.28, repeated here:

$$\mathbf{I}_{prob}^-(\mathbf{x}^+) = \int_{\mathbf{x}} p(\mathbf{x}^+|\mathbf{x}, \mathbf{u})\mathbf{I}_{prob}(\mathbf{x})d\mathbf{x} \qquad (4.1)$$

$$\mathbf{I}_{prob}^+(\mathbf{x}^+) = Cp(\mathbf{z}^+|\mathbf{x}^+, \mathbf{u})\mathbf{I}_{prob}^-(\mathbf{x}^+) \qquad (4.2)$$

where C is a normalising constant which ensures that $\mathbf{I}_{prob}^+(\mathbf{x}^+)$ integrates to one. Again, Equations 4.1 and 4.2 have a prediction-correction form familiar in robotics: the first predicts the belief forward according to the action, while the second corrects the belief using the observation. Particle-based localisers usually perform Sampling Importance Resampling (SIR) filtering [111]. This involves representing the prior with a weighted set of samples $\mathbf{Q}$. $q_j \in \mathbf{Q}$ denotes the tuple $< \mathbf{x}_j, w_j >$, where w_j is the weighting of the j'th particle. The prediction step samples a new set of predicted particles according to the process model, while the correction step re-weights each particle according to the observation likelihood function, producing $\mathbf{Q}^+$.

Applying SIR filtering to the belief transition operator f_I involves two extra steps: mapping from $\mathcal{I}_{gauss}$ to a set of particles before the update, and mapping from the particle set back to $\mathcal{I}_{gauss}$ after the update. The steps involved are:

1. sample from the parametric representation $\mathbf{I}$ to produce a set of particles $\mathbf{Q}$;

2. apply the action and observation, using Equations 4.1 and 4.2, to produce a new set of samples $\mathbf{Q}^+$; then

3. estimate the parameters of the resultant parametric representation, $\mathbf{I}^+$, from $\mathbf{Q}^+$.

The final step, of mapping back to $\mathcal{I}_{gauss}$, is clearly an approximation since $\mathbf{Q}^+$ will not in general be a true Gaussian. We will show that, at least for the problems considered in this book, the approximation is sufficiently close to the truth to generate good policies while keeping the dimensionality of the belief-space low.

Using this approach, Algorithm 7 can replace Algorithm 6 for generating Δ. Algorithm 7 requires the specification of one extra free parameter to define the number of particles representing each distribution, denoted N_Q.

Algorithm 7 consists of several distinct parts. Steps 3-5 produce a distribution over expected observations, given the action and prior belief. Steps 7-10 swap representations, producing a set of samples from the parametric representation. Steps 12-16 update the belief using SIR filtering: step 13 predicts and step 14 corrects. Finally, steps 17-18 map back to a parametric representation.

Algorithm 7 A version of **generateDistributionOverPosteriors(I, u)** which improves on Algorithm 6, by using Monte Carlo methods to allow the incorporation of different kinds of information. The outer loop produces samples in belief-space. The inner loops produce samples in state-space. The purpose of steps 3-5 is simply to sample an observation, while the purpose of steps 6-16 is to calculate the posterior belief resulting from that observation.

---

1    $\Delta \leftarrow \emptyset$
2   *foreach* $n \leftarrow 1 \dots N$
3   sample a state $\mathbf{x}$ from $p(\mathbf{x}|\mathbf{I})$
4   sample a next-state $\mathbf{x}^+$ from $p(\mathbf{x}^+|\mathbf{x}, \mathbf{u})$
5   sample an observation $\mathbf{z}^+$ from $p(\mathbf{z}^+|\mathbf{x}^+, \mathbf{u})$
6        $\mathbf{Q} \leftarrow \emptyset$
7   *foreach* $j \leftarrow 1 \dots N_Q$
8   sample a state $\mathbf{x}_j$ from $p(\mathbf{x}|\mathbf{I})$
9   add $q_j = < \mathbf{x}_j, w_j >$ to $\mathbf{Q}$, where $w_j = \frac{1}{N_Q}$
10   *end foreach j*
11       $\mathbf{Q}^+ \leftarrow \emptyset$
12   *foreach* $j \leftarrow 1 \dots N_Q$
13   sample a next-state $\mathbf{x}_j^+$ from $p(\mathbf{x}^+|\mathbf{x}_j, \mathbf{u})$
14   calculate the weight $w_j^+ = p(\mathbf{z}^+|\mathbf{x}_j^+, \mathbf{u}) w_j$
15   add $q_j^+ = < \mathbf{x}_j^+, w_j^+ >$ to $\mathbf{Q}^+$
16   *end foreach j*
17   calculate the mean μ and covariance Σ of $\mathbf{Q}^+$
18   add $\mathbf{I}^+ = < \mu, \Sigma >$ to Δ
19   *end foreach n*

The Algorithm can also be seen to consist of two nested loops: a planner which makes use of an estimator. Steps 3-5 and 17-18 involve forward prediction for planning: predicting which observations are likely to arise for the given belief and action, and seeing which belief results from each observation. This is implemented using the estimator in steps 6-16: given a known prior, action, and observation, steps 6-16 calculate the resultant posterior belief.

Algorithm 7 is very similar to the core belief projection algorithm in [108]. The major difference is that the input and output of Algorithm 7 are Gaussian distributions rather than sets of particles.

4.3.2   Stochastic Effects

One consideration for Algorithm 7 is the effect of sampling from the parametric distribution and re-estimating the parameters from those samples. Consider taking a one-dimensional Gaussian, $\mathcal{N}(\mu, \sigma^2)$, sampling a set of particles $\mathbf{Q}$ from it, then re-estimating the parameters of a new distribution, $\mathcal{N}(\mu^+, (\sigma^+)^2)$, from $\mathbf{Q}$. The new parameters, μ^+ and $(\sigma^+)^2$, can be calculated using the estimators

$$\hat{\mu}(\mathbf{Q}) = \sum w_i \mathbf{x}_i \tag{4.3}$$

and

$$\hat{\sigma}^2(\mathbf{Q}) = \frac{s}{s-1} \sum [w_i(\mathbf{x}_i - \mu^+)^2] \tag{4.4}$$

where s is the *effective sample size*, given by

$$s = \frac{1}{\sum (w^i)^2} \tag{4.5}$$

μ^+ and $(\sigma^+)^2$ can be considered to be random variables, given the non-deterministic nature of $\mathbf{Q}$. In general they are unlikely to match the originals exactly (technically the probability is zero), however normalising the variance calculation by $\frac{s}{s-1}$ ensures that they are unbiased estimates. This means that, for a number of random sample sets $\mathbf{Q}$ drawn from the original distribution, the expected values of the estimates are equal to the originals:

$$E[(\sigma^+)^2|\mu, \sigma^2] = \int_{(\sigma^+)^2} (\sigma^+)^2 p((\sigma^+)^2|\mu, \sigma^2) d(\sigma^+)^2 \tag{4.6}$$

$$= \int_{\mathbf{Q}} \hat{\sigma}^2(\mathbf{Q}) p(\mathbf{Q}|\mu, \sigma^2) d\mathbf{Q} \tag{4.7}$$

$$= \sigma^2 \tag{4.8}$$

and similarly,

$$E[\mu^+|\mu, \sigma^2] = \mu \tag{4.9}$$

However, while the estimates are unbiased, the distribution $p((\sigma^+)^2|\mu, \sigma^2)$ is skew[1]. Figure 4.5 shows the distributions $p((\sigma^+)^2|\mu, \sigma^2)$ and $p(\mu^+|\mu, \sigma^2)$, evaluated numerically by conditioning on $\mathbf{Q}$ as in Equation 4.7. Since $p((\sigma^+)^2|\mu, \sigma^2)$ has a longer tail to the right, more of the probability mass must lie below σ^2 than above in order for Equation 4.8 to hold. In other words, it is more likely that $(\sigma^+)^2$ will under-estimate σ^2 than over-estimate it, but over-estimates are likely to be further from the truth.

From the planner's point of view, the estimator is used sequentially. That is, at each time step a distribution is sampled, the samples are possibly modified, then a new distribution is re-estimated and used as the input to the next iteration. Assuming no modification to the samples, the symmetry of $p(\mu^+|\mu, \sigma^2)$ means that, while μ^+ undergoes a random walk during repeated sequential re-estimations, the result is as likely to be too small as too large. The skewness of $p((\sigma^+)^2|\mu, \sigma^2)$, however, means that the estimate of the variance is likely to become gradually smaller. This is demonstrated in the simulation shown in Figure 4.6. At each iteration of the simulation 200 samples are drawn from a Gaussian distribution. A new Gaussian is then re-estimated from those samples to provide the input for the next iteration. The simulation begins with the distribution $N(0, 1)$.

---

[1]For a discussion of the skewness of the likelihood $p(\mathbf{Q}|\sigma^2)$, see Section 24.1 of [69].

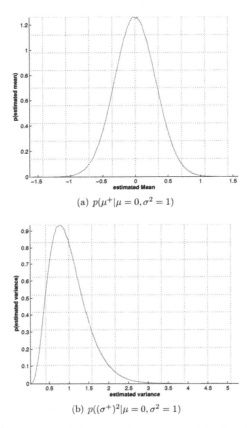

(a) $p(\mu^+|\mu = 0, \sigma^2 = 1)$

(b) $p((\sigma^+)^2|\mu = 0, \sigma^2 = 1)$

Figure 4.5: The distributions over (a) estimated mean, and (b) estimated variance, given the mean and variance from the previous iteration. Distributions are evaluated by drawing 2×10^6 particle sets of size 10 from $\mathcal{N}(0, 1)$. Only 10 samples were used in order to make the skewness of $p((\sigma^+)^2|\mu, \sigma^2)$ clear. 56% of the probability mass of $p((\sigma^+)^2|\mu, \sigma^2)$ lies below σ^2. $p(\mu^+|\mu, \sigma^2)$ is symmetric.

In addition to the average behaviour of the estimator, the variation in its behaviour is of concern. If Algorithm 7 leads the planner to believe that certain action sequences are likely to reduce its uncertainty, it has no way of knowing whether this is really the case, or whether this is simply an artefact of the process of sampling and re-estimating. It will produce bad plans, attempting to take advantage of the expected but unattainable reduction in uncertainty.

In an attempt to combat uncertainty under-estimates in both the average and worst case, we multiply the estimated covariance matrix by the factor ω. For the BlockWorld experiments, $\omega = 1.1$ was used. This value is clearly too large to offset the fact that uncertainty decreases

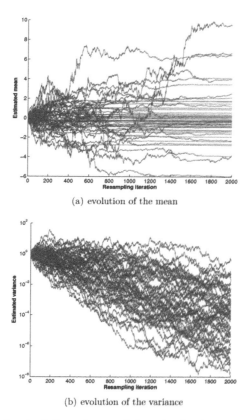

(a) evolution of the mean

(b) evolution of the variance

Figure 4.6: 50 trials of a simulation where each iteration involves sampling from a Gaussian distribution then re-estimating new parameters from those samples, showing the evolution of the estimate of (a) the mean and (b) the variance. A sample size of 200 was used. The estimated variance tends to drift downwards. Starting from a variance of 1, after 2000 iterations the mean estimated variance is 0.0078. The stabilisation of the estimate of the mean is due to the fact that the variance is changing simultaneously. As the variance becomes smaller, the mean becomes more stable. Note that the y-axis of plot (a) is linear, while the y-axis of plot (b) is logarithmic.

in the average case, and too small to correct the absolute worst case, however it was found empirically to produce reasonable results. The severity of these problems decreases with the size of the sample set. Chapter 8 will discuss the use of PPOMDP in the application domain introduced in Chapter 1. Since more samples are used for this more realistic problem, a factor of $\omega = 1.0$ was used.

4.4 Efficient Calculation of the Belief Transition Function

The previous section presented an algorithm for transitioning beliefs, capable of incorporating all available information. This section improves the efficiency of that algorithm by re-using predictions and reducing likelihood calculations.

4.4.1 Re-use of Predictions

Algorithm 7 can be calculated more efficiently by noting that the process model $p(\mathbf{x}^+|\mathbf{x},\mathbf{u})$ is invoked in two places. The first (step 4) generates a predicted distribution from which to sample observations. The second (step 12) generates a predicted distribution which is re-weighted by an observation to produce a posterior.

Instead, the predicted distribution used to generate observations can be re-used for generating posteriors. Algorithm 8 modifies Algorithm 7 by using this idea. Note that this introduces the requirement that the number of observation samples, and hence the number of samples used to represent distributions over posteriors, be equal to the number of state samples used to represent each posterior. If N denotes this number, the total number of particle predictions in Algorithm 8 is N, compared with $N(1 + N_Q)$ in Algorithm 7.

Algorithm 8 A version of **generateDistributionOverPosteriors(I, u)** which extends Algorithm 7 by re-using predictions. Steps 2-6 map the prior from a parametric representation to a set of samples, predict those samples forward, and generate a distribution over observations. Steps 9-14 calculate the posterior, in parametric form, resulting from each observation.

---
1   $\Delta \leftarrow \emptyset$
2   *foreach* $i \in 1 \ldots N$
3   sample a state $\mathbf{x}_i$ from $p(\mathbf{x}|\mathbf{I})$
4   sample a next-state $\mathbf{x}_i^+$ from $p(\mathbf{x}^+|\mathbf{x}_i,\mathbf{u})$
5   sample an observation $\mathbf{z}_i^+$ from $p(\mathbf{z}^+|\mathbf{x}_i^+,\mathbf{u})$
6   *end foreach* i
7   *foreach* $i \in 1 \ldots N$
8        $\mathbf{Q}^+ \leftarrow \emptyset$
9   *foreach* $j \in 1 \ldots N$
10   calculate the likelihood $w_j^+ = p(\mathbf{z}_i^+|\mathbf{x}_j^+,\mathbf{u})$
11   add $q_j^+ = <\mathbf{x}_j^+, w_j^+>$ to $\mathbf{Q}^+$
12   *end foreach* j
13   calculate the mean μ and covariance Σ of $\mathbf{Q}^+$
14   add $\mathbf{I}^+ = <\mu, \Sigma>$ to Δ
15 *end foreach* i

To help clarify Algorithm 8, Figure 4.7 shows a simple example. An agent is navigating in a one-dimensional world. It can localise by sensing the distance to the wall to its right. The prior at time k is shown in Figure 4.7(a). Steps 3 and 4 of Algorithm 8 sample from that prior (using only four samples) and predict according to the process model, as shown in Figures 4.7(b) and (c) respectively. In step 5, noisy observations of the range to the wall are sampled from the observation model, as depicted in Figure 4.7(d). The set of sampled observations is an approximation to the distribution over observations. For each observation, steps 10-11 (Figure 4.7(e)) calculate an associated posterior by weighting the particles according to the likelihood function. Finally, step 13 maps the posterior back into parametric form, as shown in Figure 4.7(f).

Note that re-using the prediction in this way is likely to have a small effect on the resultant posteriors. Strictly speaking, the set of particles used to generate the distribution over observations and the set of particles used to calculate the posterior resulting from an observation should be independent. Using the same set of particles for both enforces a relation between one particle and each observation. This effect can be avoided by omitting the particle that generated the observation from the set of particles representing that posterior. In practice however, for a large enough set of particles the effect should be negligible.

4.4.2   Reducing Likelihood Calculations

Steps 5 and 10 in Algorithm 8, which sample observations and calculate likelihoods respectively, are particularly computationally demanding. An observation sample z_i^+ is usually drawn from $p(z_i^+|x_i^+, u)$ by calculating the expected observation from state x_i^+, given action u, and perturbing it according to a sensor model. Calculation of the expected observation is potentially expensive. For example, when localising using a range-bearing sensor in an occupancy grid, this calculation requires a costly ray-trace through that occupancy grid to find the expected ranges.

The cost of each likelihood calculation (step 10) is also high. The likelihood of making observation z_i^+ from state x_j^+, given action u, is denoted $p(z_i^+|x_j^+, u)$. It is usually approximated by $p(z_i^+|\hat{z}_j^+, u)$, the likelihood of observing z_i^+ based on $\hat{z}_j^+$, the expected observation from state x_j^+ given action u [44]. To take the occupancy grid example again, $p(z_i^+|x_j^+, u)$ can be obtained by using ray-tracing to compute the expected range $\hat{z}_j^+$, then assuming a Gaussian sensor noise model for $p(z_i^+|\hat{z}_j^+, u)$, as shown in Figure 4.8. This approximation allows the likelihood function to be defined purely in observation-space, independently from the state. It requires the calculation of an expected observation followed by a comparison of observations.

Algorithm 8 therefore calculates a total of $N^2 + N$ expected observations, plus N^2 observation comparisons. Step 5 requires N observation calculations (one per particle). Since step 10 is inside a double-loop, it requires N^2 observation calculations and observation comparisons. The computational cost of this is potentially crippling. Under certain assumptions it can be reduced

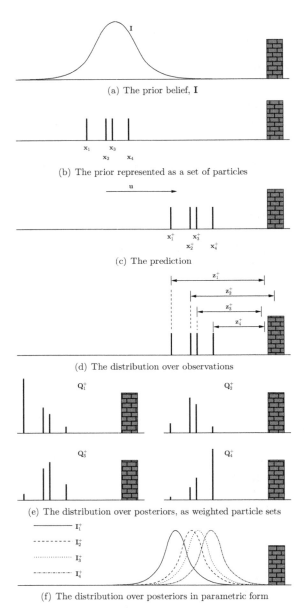

(a) The prior belief, **I**

(b) The prior represented as a set of particles

(c) The prediction

(d) The distribution over observations

(e) The distribution over posteriors, as weighted particle sets

(f) The distribution over posteriors in parametric form

Figure 4.7: A simple one-dimensional example illustrating the mechanics of Algorithm 8 with only four particles.

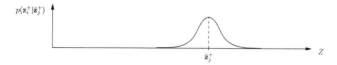

Figure 4.8: $p(\mathbf{z}_i^+|\mathbf{x}_j^+, \mathbf{u})$ is usually approximated by $p(\mathbf{z}_i^+|\hat{\mathbf{z}}_j^+, \mathbf{u})$: the likelihood of observing $\mathbf{z}_i^+$ based on the expected observation from state $\mathbf{x}_j^+$, denoted $\hat{\mathbf{z}}_j^+$. A Gaussian sensor noise model is often assumed for $p(\mathbf{z}_i^+|\hat{\mathbf{z}}_j^+, \mathbf{u})$, as shown here.

in two ways: (a) by reducing the number of required observation calculations, and (b) by using symmetry to reduce the number of likelihood calculations.

Reducing Observation Calculations

Let W_{ij} denote the weighting of the j'th particle in the i'th posterior. It is proportional to the likelihood $L_{ij} = p(\mathbf{z}_i^+|\hat{\mathbf{z}}_j^+, \mathbf{u})$. Under the assumption that the uncertainty in the observation is small relative to the uncertainty in the state distribution, one can make the approximation that

$$L_{ij} = p(\mathbf{z}_i^+|\hat{\mathbf{z}}_j^+, \mathbf{u}) \qquad (4.10)$$

$$\simeq p(\hat{\mathbf{z}}_i^+|\hat{\mathbf{z}}_j^+, \mathbf{u}) \qquad (4.11)$$

That is, the likelihood of the j'th particle given an observation sampled from state $\mathbf{x}_i^+$ is approximately equal to the likelihood given the expected observation from state $\mathbf{x}_i^+$. Let this approximate likelihood be denoted $\tilde{L}_{ij}$. The accuracy of the approximation relies on the state uncertainty being large relative to the sensor noise, as is often the case for example when using laser range finders for mobile robot localisation. This approximation was used successfully for planning by Roy [92].

Given this assumption, rather than computing $N^2 + N$ observations, only N need be computed: the expected observation from each predicted state sample. The weighting of the j'th particle in the i'th posterior is then given by

$$W_{ij} = \frac{1}{C}\tilde{L}_{ij} \qquad (4.12)$$

where $C = \sum_{j=1}^{N} \tilde{L}_{ij}$ is a normalising factor.

It can be helpful to think of W as a weight matrix, as illustrated in Figure 4.9: the i'th observation corresponds to a row of W, where the j'th element of that row gives the weighting of the j'th particle for the posterior resulting from that observation. W can be constructed by calculating $\tilde{L}_{ij}$ for all i and j, then normalising each row. As stated in Section 4.4.1, the particle that generated an observation should technically be omitted when calculating the posterior

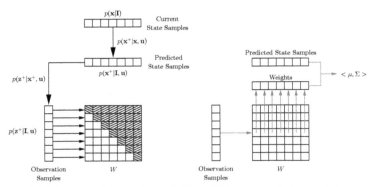

(a) Calculation of the weight matrix W (b) W is used to calculate the i'th possible posterior (resulting from the i'th observation).

Figure 4.9: A graphical depiction of the operation of Algorithm 9. (a) shows the generation of the weight matrix W. From a set of samples of the current belief, a set of predicted samples and a distribution over future observations is generated. Under the symmetry condition described in Section 4.4.2, only the upper-triangular (shaded) entries in the weight matrix need to be calculated by comparing observations. These values can then be copied to the lower triangle. Finally, each row of the matrix is normalised. The entries on the main diagonal should technically be ignored, however for implementational simplicity we do not. (b) To calculate the posterior resulting from the i'th observation, the i'th row of W is used to weight the predicted particles. Calculating the statistics of the weighted particles produces the parametric posterior.

corresponding to that observation. This corresponds to omitting the diagonal elements of W. For simplicity of implementation however, the results presented in this work were generated without omitting the diagonal elements.

Exploiting Likelihood Function Symmetry

In addition to the reduction in the number of observations which must be computed, the number of likelihood calculations can be reduced under some circumstances. In general, calculating the elements of W requires N^2 observation comparisons. However if the matrix is symmetric then only the upper-triangular portion needs to be calculated. This will be the case when the likelihood function is symmetric, namely

$$p(z^i|z^j, \mathbf{u}) = p(z^j|z^i, \mathbf{u}) \tag{4.13}$$

When this condition holds,

$$\tilde{L}_{ij} = p(\hat{\mathbf{z}}_i^+|\hat{\mathbf{z}}_j^+, \mathbf{u}) \tag{4.14}$$

$$= p(\hat{\mathbf{z}}_j^+|\hat{\mathbf{z}}_i^+, \mathbf{u}) \tag{4.15}$$

$$= \tilde{L}_{ji} \tag{4.16}$$

where the first and third lines follow from the definition of $\tilde{L}$, and the second from Equation 4.13. The assumption of likelihood function symmetry is often made for robot localisation problems. In this case it approximately halves the required number of likelihood calculations.

Algorithm 9 describes the addition of these two optimisations. To try to help develop intuition for the problem, Figure 4.9 illustrates the operation of Algorithm 9 graphically.

Algorithm 9 A version of **generateDistributionOverPosteriors(I, u)** which optimises Algorithm 8 by re-using the calculation of expected observations and likelihoods.

```
1    Δ ← ∅
2    foreach i ∈ 1 . . . N
3        sample a state x_i from p(x|I)
4        sample a next-state x_i⁺ from p(x⁺|x_i, u)
5        calculate the expected observation ẑ_i⁺ from state x_i⁺, given u
6    end foreach i
7    foreach i ∈ 1 . . . N
8        foreach j ∈ i . . . N
9            calculate L̃_ij = p(ẑ_i⁺, ẑ_j⁺)
10       end foreach j
11   end foreach i
12   copy the upper triangle of L̃ to the lower triangle
13   normalise each row of L̃ to produce W
14   foreach i ∈ 1 . . . N
15       Q⁺ ← ∅
16       foreach j ∈ 1 . . . N
17           add q_j⁺ =< x_j⁺, W_ij > to Q⁺
18       end foreach j
19       calculate the mean μ and covariance Σ of Q⁺
20       add I⁺ =< μ, Σ > to Δ
21   end foreach i
```

4.5   Scalability

Algorithm 9 is executed once for every belief-action combination, and makes a nested loop over samples. Ignoring the cost of the weighting function and updates to T for simplicity, the cost of the particle-based PPOMDP algorithm is therefore

$$O\bigg(|B||U|N\big(C(\mathbf{z}^+) + NC(L)\big)\bigg) \tag{4.17}$$

where $C(\mathbf{z}^+)$ and $C(L)$ are the costs of calculating expected observations and likelihoods, respectively. Ignoring these constants, the cost is given by

$$O(|B||U|N^2) \tag{4.18}$$

There are strong parallels between the algorithm presented in this chapter and Roy's Belief Compression and AMDP algorithms [92]. These algorithms operate in the space of restricted classes of distributions over discrete states. In order to transition beliefs, they map to that underlying discrete space, transition the belief there, then map back. Our approach is similar: we operate in a restricted class of beliefs. In order to transition we map to the space of particles, perform the transition, then map back.

An important distinction is that the particle-based PPOMDP algorithm does not rely on an underlying discrete representation. The cost of transitioning a belief over discrete states, for a given prior, action and observation, scales quadratically with the number of discrete states (although this can be minimised using sparse matrix methods). Assuming the observation space is represented with N samples, methods based on FVI and an underlying discrete state-space therefore scale according to

$$O(|B||U|N|S|^2) \tag{4.19}$$

The advantage of the algorithm advocated in this book is that the cost of each transition is independent of the physical size of the environment. We will show in subsequent chapters that this allows PPOMDP to scale to complex real-world problems such as the one introduced in Chapter 1. The distinction is related to the difference between grid-based Markov localisation [79][98][21] and particle filters [109]. The former must spread computation evenly over the entire state-space, regardless of which areas are more relevant. This approach has problems scaling to larger physical environments, because it must constantly update areas of the state-space which are relatively unlikely. Particle filters have been more successful. A particle filter can use a high density of particles in areas of relevance, producing accurate results. To attain similar accuracy, a grid-based Markov localisation scheme must use a fine grid over the entire state-space, which cannot scale. Furthermore, a particle filter need not choose a constant number of particles. For example, a KLD particle filter [42] selects the number of particles based on the uncertainty of the distribution. While this extension has not been implemented, a similar approach applied to the PPOMDP belief transition function would further improve accuracy and scalability.

4.6   Results

The particle-based PPOMDP algorithm described by Algorithm 9 was compared against the three previously-compared algorithms on the worlds from Section 4.1. The particle-based version used the same parameters as the basic version. The only additional free parameter is the number of samples, which was set to 50. Informal experiments showed that increasing this number had little effect on performance.

As in the previous chapter, the estimator for belief tracking uses the same belief update function as the planner. In other words, the belief tracker maps from a Gaussian to particles and back again at every iteration. This is clearly unnecessary and will adversely affect the accuracy of the belief tracker. A better approach would be to use a standard particle filter, mapping to a Gaussian at each iteration solely in order to generate a Gaussian belief from which the policy can produce an action. If such an approach were implemented, it would be likely to improve results.

The results are shown in Figure 4.10. The particle-based PPOMDP algorithm clearly produces excellent results on all four worlds, while taking a similar length of time to converge as the basic PPOMDP algorithm. Furthermore, the dysfunctional behaviours identified at the beginning of this chapter seem to have disappeared. When the agent is behind an obstacle, an action which causes a collision results in the belief being updated, such that the agent tries a different action on the subsequent time step. The behaviour of repeating endless cycles also seems to have gone. Instead, the agent appears to commit to a strategy. Since the particle-based PPOMDP algorithm outperforms PERSEUS on this world, we cease comparing against PERSEUS for improvements discussed in later chapters.

4.7   Summary

This chapter presented an approach to constructing an accurate and efficient belief transition function, capable of incorporating information (such as negative information) which would produce non-Gaussian posteriors. After demonstrating and analysing the need to incorporate such information, a basic approach based on Monte Carlo methods was presented. A number of optimisations for this approach were then described, in order to ensure its suitability for practical application. The improved algorithm was empirically evaluated on several different versions of BlockWorld. When compared against MDP, PERSEUS and Basic-PPOMDP, it demonstrated excellent performance, with low computational requirements similar to Basic-PPOMDP's. The scalability of the algorithm was then analysed and discussed, showing how the particle-based PPOMDP algorithm has fundamental differences which make it better able to scale to large environments than algorithms with underlying discrete state-spaces.

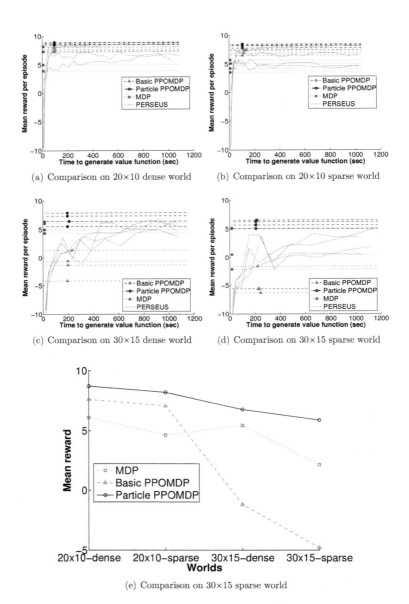

(a) Comparison on 20×10 dense world   (b) Comparison on 20×10 sparse world

(c) Comparison on 30×15 dense world   (d) Comparison on 30×15 sparse world

(e) Comparison on 30×15 sparse world

Figure 4.10: (a-d) show the mean reward per episode for each goal configuration on each world. Each datapoint in (a-d) is the average of 1000 episodes. (e) shows the mean reward per episode, averaged over goal locations, for each world.

While the particle-based PPOMDP algorithm produces good results, it still requires a significant amount of planning time on the larger of the toy worlds. In its current form, it is unlikely
to scale to a more realistic problem. The following chapter presents an approach to improving
the algorithm's scalability.

Chapter 5

Factoring Observations

While the algorithm presented in the previous chapter is capable of generating good plans, the time required for planning is still considerable. While adequate for BlockWorld, it is unlikely to be directly applicable to more realistic scenarios.

The POMDP problem can be represented using a graphical model, as was shown in Section 2.7.5. Figure 5.1 reproduces a single time slice of that model. In general, the sizes of the conditional probability tables (CPTs) in graphical models are determined by the number of parents of each node, called their fan-in [83]. The bottleneck for the PPOMDP algorithm is the node labelled I^+. The fan-in of this node results in a large CPT which is expensive to calculate, especially for high-dimensional observations

This chapter describes how the bottleneck can be alleviated. Conditional independence assumptions can be exploited by adding extra states, allowing a factoring of the problem which permits the partial effects of observations to be pre-calculated. With certain approximations, this novel approach allows algorithms based on fitted value iteration to be broken into smaller components, reducing the total computational complexity.

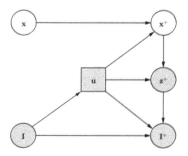

Figure 5.1: Graphical model showing one update of the POMDP.

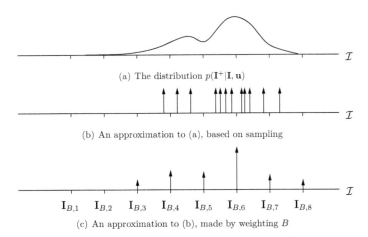

(a) The distribution $p(\mathbf{I}^+|\mathbf{I}, \mathbf{u})$

(b) An approximation to (a), based on sampling

$I_{B,1}$ $I_{B,2}$ $I_{B,3}$ $I_{B,4}$ $I_{B,5}$ $I_{B,6}$ $I_{B,7}$ $I_{B,8}$

(c) An approximation to (b), made by weighting B

Figure 5.2: An example of a hypothetical one-dimensional belief-space, showing approximations to the continuous distribution over posteriors $p(\mathbf{I}^+|\mathbf{I}, \mathbf{u})$ in (a). (b) illustrates $\sum_{\mathbf{z}^+} p(\mathbf{z}^+|\mathbf{I}, \mathbf{u})p(\mathbf{I}^+|\mathbf{I}, \mathbf{u}, \mathbf{z}^+)$, an approximation to (a) based on sampling observations. In this case only 12 samples are drawn. The heights of the samples show their weights (which are uniform in (b)). (c) illustrates an approximation to (b) based on weighting the set of beliefs B. In this case linear interpolation is used. Each sample in (b) induces a weight for the two nearby beliefs in B. The summation of these weights produces (c).

The remainder of this chapter proceeds as follows. Section 5.1 presents a slightly different view of fitted value iteration, and Section 5.2 shows how the POMDP problem can be factored. Section 5.3 brings the two previous sections together, showing how this factoring can be exploited using the view of FVI presented in Section 5.1. A modified PPOMDP solution algorithm, based on these insights, is described in Section 5.4. Section 5.5 applies the modified solution algorithm to the BlockWorld problem, Section 5.6 presents results which show a significant improvement in planning speed, and Section 5.7 concludes.

5.1   A Different View of the FVI Approximation

For each belief and action, Algorithm 4 from the previous chapter performs two steps. It first calculates a distribution over posteriors, then uses this distribution, together with a weighting function, to calculate transition probabilities between discrete beliefs in B. This section shows how this can be viewed as approximating a continuous distribution by a set of Dirac delta functions. This view will be used subsequently to improve the efficiency of the PPOMDP algorithm.

This discussion assumes a continuous observation space. When considering a particular action

$\mathbf{u}$ from a particular belief $\mathbf{I}$, the agent is aware of a continuous distribution over possible next-I-states, as illustrated in Figure 5.2(a). This distribution is denoted $p(\mathbf{I}^+|\mathbf{I}, \mathbf{u})$ and is obtained by integrating over all possible next-observations $\mathbf{z}^+$. The integral can be approximated by sampling from the observation space (Section 3.4.1 showed how to perform this sampling efficiently, by conditioning on the current belief). Given a discrete set of observation samples, $p(\mathbf{I}^+|\mathbf{I}, \mathbf{u})$ can be approximated by a set of Dirac delta functions:

$$p(\mathbf{I}^+|\mathbf{I}, \mathbf{u}) \simeq \sum_{\mathbf{z}^+} p(\mathbf{z}^+|\mathbf{I}, \mathbf{u}) p(\mathbf{I}^+|\mathbf{I}, \mathbf{u}, \mathbf{z}^+) \tag{5.1}$$

$$= \sum_{\mathbf{z}^+} p(\mathbf{z}^+|\mathbf{I}, \mathbf{u}) \delta(f_{\mathbf{I}}(\mathbf{I}, \mathbf{u}, \mathbf{z}^+)) \tag{5.2}$$

as illustrated in Figure 5.2(b).

When evaluating the merit of taking the action $\mathbf{u}$, the agent must know the value of each next-I-state on which a delta is centred. The problem is that, since the I-space is continuous, one cannot store the values of all possible next-I-states. As described in Section 2.5.2, the solution offered by fitted value iteration can be seen as approximating $p(\mathbf{I}^+|\mathbf{I}, \mathbf{u}, \mathbf{z}^+)$ by a mixture of delta functions centred on a set of I-states B whose values are explicitly stored:

$$p(\mathbf{I}^+|\mathbf{I}, \mathbf{u}, \mathbf{z}^+) \simeq \sum_{j=1}^{|B|} \lambda_B\big(f_{\mathbf{I}}(\mathbf{I}, \mathbf{u}, \mathbf{z}^+), j\big) \delta(\mathbf{I}_{B,j}^+) \tag{5.3}$$

where λ_B is a weighting function defined over the set B, as used in previous chapters. λ_B interpolates the I-state $\mathbf{I}^+ = f_{\mathbf{I}}(\mathbf{I}, \mathbf{u}, \mathbf{z}^+)$ onto nearby beliefs in B. This approximation is illustrated in Figure 5.2(c).

Substituting Equation 5.3 into Equation 5.1, and assuming that the current I-state is the i'th member of B, denoted $\mathbf{I}_{B,i}$, gives

$$p(\mathbf{I}^+|\mathbf{I}_{B,i}, \mathbf{u}) \simeq \sum_{\mathbf{z}^+} p(\mathbf{z}^+|\mathbf{I}_{B,i}, \mathbf{u}) \sum_{j=1}^{|B|} \lambda_B\big(f_{\mathbf{I}}(\mathbf{I}_{B,i}, \mathbf{u}, \mathbf{z}^+), j\big) \delta(\mathbf{I}_{B,j}^+) \tag{5.4}$$

This expresses the transition probabilities from $\mathbf{I}_{B,i}$ purely in terms of future beliefs in B. The transition probability from the i'th to the j'th belief in B, denoted $T(\mathbf{I}_{B,i}, \mathbf{u}, \mathbf{I}_{B,j}^+)$, is

$$T(\mathbf{I}_{B,i}, \mathbf{u}, \mathbf{I}_{B,j}^+) = \sum_{\mathbf{z}^+} p(\mathbf{z}^+|\mathbf{I}_{B,i}, \mathbf{u}) \lambda_B\big(f_{\mathbf{I}}(\mathbf{I}_{B,i}, \mathbf{u}, \mathbf{z}^+), j\big) \tag{5.5}$$

which is identical to Equation 3.3. Equation 5.4 defines a distribution over next-I-states from $\mathbf{I}_{B,i}$, and Equation 5.5 gives the value of that distribution for the next-I-state $\mathbf{I}_{B,j}^+$.

The previous chapter showed how the set of transition probabilities T could be calculated using

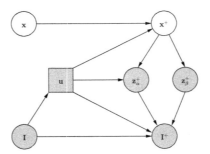

Figure 5.3: Graphical model showing the observation split into two conditionally-independent components, $\mathbf{z}_\alpha^+$ and $\mathbf{z}_\beta^+$, one of which is also independent from the action.

a nested sum over all possible combinations of $\mathbf{I}^B, \mathbf{u}$ and $\mathbf{z}^+$. This chapter explores approaches to mitigating the cost of this nested sum.

5.2   Factoring Conditionally-Independent Observation Components

The computational burden can be reduced if $p(\mathbf{I}^+|\mathbf{I}, \mathbf{u}, \mathbf{z}^+)$ can be factored in some way based on conditional independence assumptions. Suppose the observation vector can be split into components, such that $\mathbf{z}^+ = \{\mathbf{z}_\alpha^+, \mathbf{z}_\beta^+, \mathbf{z}_\gamma^+, \dots\}$, where all components are conditionally independent given the state. Furthermore, suppose that some subset is conditionally independent of the action given the state. The particular manner in which the distribution can be factored depends on the specifics of the problem, however the conditional independence assumption is not unreasonable. It is often assumed in robot navigation and data fusion problems that individual sensors are conditionally independent from each other and from the action, given the state [35][113]. The remainder of this discussion will consider two components, $\mathbf{z}_\alpha^+ \in Z_\alpha$ and $\mathbf{z}_\beta^+ \in Z_\beta$, only one of which is action-dependent, as shown in Figure 5.3.

5.2.1   Conditioning on the Belief after Acting and Partial Observation

To simplify the calculation of $p(\mathbf{I}^+|\mathbf{I}, \mathbf{u})$, Figure 5.4 postulates an extra I-state $\mathbf{I}_\alpha^+ \in \mathcal{I}_\alpha$, representing the belief after acting and observing $\mathbf{z}_\alpha^+$. Technically, $\mathcal{I}_\alpha$ is not in the same I-space as $\mathbf{I}$ or $\mathbf{I}^+$, which are in a space derived from $\mathcal{I}_{hist}$: the space of histories of actions and complete observations. When dealing with $\mathcal{I}_{prob}$ however, the space is the same: both are distributions over state-space.

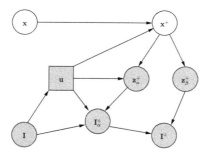

Figure 5.4: A factored version of Figure 5.3, postulating an intermediate state, $\mathbf{I}_\alpha^+$, representing the belief after acting and incorporating the observation component $\mathbf{z}_\alpha^+$.

Let $f_{\mathbf{I}_\alpha}$ and $f_{\mathbf{I}_\beta}$ denote the deterministic functions mapping to and from $\mathbf{I}_\alpha^+$:

$$\mathbf{I}_\alpha^+ = f_{\mathbf{I}_\alpha}(\mathbf{I}, \mathbf{u}, \mathbf{z}_\alpha^+) \tag{5.6}$$

$$\mathbf{I}^+ = f_{\mathbf{I}_\beta}(\mathbf{I}_\alpha^+, \mathbf{z}_\beta^+) \tag{5.7}$$

The following probability distributions can then be defined:

$$p(\mathbf{I}_\alpha^+ | \mathbf{I}, \mathbf{u}, \mathbf{z}_\alpha^+) = \delta(f_{\mathbf{I}_\alpha}(\mathbf{I}, \mathbf{u}, \mathbf{z}_\alpha^+)) \tag{5.8}$$

$$p(\mathbf{I}^+ | \mathbf{I}_\alpha^+, \mathbf{z}_\beta^+) = \delta(f_{\mathbf{I}_\beta}(\mathbf{I}_\alpha^+, \mathbf{z}_\beta^+)) \tag{5.9}$$

These distributions can be used to expand $p(\mathbf{I}^+ | \mathbf{I}, \mathbf{u})$ according to the total probability theorem:

$$p(\mathbf{I}^+ | \mathbf{I}, \mathbf{u}) = \int_{\mathbf{I}_\alpha^+ \in \mathcal{I}_\alpha} p(\mathbf{I}^+ | \mathbf{I}_\alpha^+) p(\mathbf{I}_\alpha^+ | \mathbf{I}, \mathbf{u}) d\mathbf{I}_\alpha^+ \tag{5.10}$$

where

$$p(\mathbf{I}^+ | \mathbf{I}_\alpha^+) = \sum_{\mathbf{z}_\beta^+} p(\mathbf{I}^+ | \mathbf{I}_\alpha^+, \mathbf{z}_\beta^+) p(\mathbf{z}_\beta^+ | \mathbf{I}_\alpha^+) \tag{5.11}$$

$$p(\mathbf{I}_\alpha^+ | \mathbf{I}, \mathbf{u}) = \sum_{\mathbf{z}_\alpha^+} p(\mathbf{I}_\alpha^+ | \mathbf{I}, \mathbf{u}, \mathbf{z}_\alpha^+) p(\mathbf{z}_\alpha^+ | \mathbf{I}, \mathbf{u}) \tag{5.12}$$

In words, the probability of getting from a particular current belief-state $\mathbf{I}$ to a particular future belief-state $\mathbf{I}^+$ is the sum of all the ways of getting there via various mid-points in $\mathcal{I}_\alpha$.

5.3   Application to the PPOMDP Solution Algorithm

The extra state can potentially be used to reduce the amount of computation required to calculate T. Compare Figures 5.3 and 5.4. Node $\mathbf{I}^+$ in Figure 5.3 has four parents. This relatively large fan-in is replaced in Figure 5.4 with fan-ins of three (to $\mathbf{I}_\alpha^+$) and two (to $\mathbf{I}^+$).

Exact marginalisation over any particular parent involves a nested iteration over all parents. Marginalising over $\mathbf{z}_\alpha^+$ and $\mathbf{z}_\beta^+$ to produce $p(\mathbf{I}^+|\mathbf{I}, \mathbf{u})$, without using $\mathbf{I}_\alpha^+$, therefore requires a giant nested loop over four variables. In contrast, $p(\mathbf{I}^+|\mathbf{I}_\alpha^+)$ can be computed in a nested loop over only two variables. This result can then be used to calculate $p(\mathbf{I}^+|\mathbf{I}, \mathbf{u})$ in a second nested loop over three variables. As will be shown, the two loops over fewer variables can be evaluated much more quickly than the single loop over many variables.

5.3.1   Representing $p(\mathbf{I}^+|\mathbf{I}_\alpha^+)$

The difficulty in representing $p(\mathbf{I}^+|\mathbf{I}_\alpha^+)$ is that the I-space is continuous, and hence $p(\mathbf{I}^+|\mathbf{I}_\alpha^+)$ cannot be stored exactly in a lookup table. However it was shown in Section 5.1 how the continuous conditional probability distribution (CPD) $p(\mathbf{I}^+|\mathbf{I}, \mathbf{u})$ can be approximated by the discrete CPT $T(\mathbf{I}_{B,i}, \mathbf{u}, \mathbf{I}_{B,j}^+)$. The same approximation method can be used for $p(\mathbf{I}^+|\mathbf{I}_\alpha^+)$.

Let $B_\alpha = \{\mathbf{I}_{B_\alpha,1}^+, \mathbf{I}_{B_\alpha,2}^+, \ldots, \mathbf{I}_{B_\alpha,|B_\alpha|}^+\}$ be a belief set of possible values for $\mathbf{I}_\alpha^+$, distinct from $B = \{\mathbf{I}_{B,1}^+, \mathbf{I}_{B,2}^+, \ldots, \mathbf{I}_{B,|B|}^+\}$ which is a set of possible values of $\mathbf{I}^+$. As pointed out in Section 5.2.1, the probability distribution $p(\mathbf{I}^+|\mathbf{I}_\alpha^+, \mathbf{z}_\beta^+)$, is a delta function centred on $f_{\mathbf{I}_\beta}(\mathbf{I}_\alpha^+, \mathbf{z}_\beta^+)$ (Equation 5.9). This can be approximated by a mixture of deltas centred on B, in analogy to Equation 5.3:

$$p(\mathbf{I}^+|\mathbf{I}_\alpha^+, \mathbf{z}_\beta^+) \simeq \sum_{i=1}^{|B|} \lambda_B\big(f_{\mathbf{I}_\beta}(\mathbf{I}_\alpha^+, \mathbf{z}_\beta^+), i\big)\delta(\mathbf{I}_{B,i}^+) \tag{5.13}$$

Substituting this approximation into Equation 5.11 gives

$$p(\mathbf{I}^+|\mathbf{I}_\alpha^+) \simeq \sum_{\mathbf{z}_\beta^+} p(\mathbf{z}_\beta^+|\mathbf{I}_\alpha^+) \sum_{i=1}^{|B|} \lambda_B\big(f_{\mathbf{I}_\beta}(\mathbf{I}_\alpha^+, \mathbf{z}_\beta^+), i\big)\delta(\mathbf{I}_{B,i}^+) \tag{5.14}$$

which can be stored explicitly as the CPT $T_\beta(\mathbf{I}_{B_\alpha,i}^+, \mathbf{I}_{B,j}^+)$. As with the approximation made by fitted value iteration, the approximation made by Equation 5.13 is unlikely to adversely affect the quality of value iteration if the value function is sufficiently smooth and the set B is sufficiently dense.

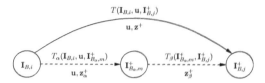

Figure 5.5: The relationship between the discrete elements used in Algorithm 10. T is approximated by T_α and T_β. The additional information required for each transition is below the arcs, the CPT describing the transition is above.

Stepping backwards through the graphical model, $p(\mathbf{I}_\alpha^+|\mathbf{I}, \mathbf{u})$ can be similarly approximated by

$$p(\mathbf{I}_\alpha^+|\mathbf{I}, \mathbf{u}) \simeq \sum_{\mathbf{z}_\alpha^+} p(\mathbf{z}_\alpha^+|\mathbf{I}, \mathbf{u}) \sum_{i=1}^{|B_\alpha|} \lambda_{B_\alpha}\left(f_{\mathbf{I}_\alpha}(\mathbf{I}, \mathbf{u}, \mathbf{z}_\alpha^+), i\right)\delta(\mathbf{I}_{B_\alpha,i}^+) \qquad (5.15)$$

where λ_{B_α} is a weighting function defined over B_α and $f_{\mathbf{I}_\alpha}$ is the function governing the belief transition based on the action and observation component $\mathbf{z}_\alpha^+$. Again, this can be stored explicitly in the CPT $T_\alpha(\mathbf{I}_{B,i}, \mathbf{u}, \mathbf{I}_{B_\alpha,j}^+)$. To clarify, Figure 5.5 shows the relationship between the various discrete I-states.

5.4 A Modified PPOMDP Solution Algorithm

Putting the approximations together gives

$$p(\mathbf{I}_{B,i}^+|\mathbf{I}_{B,j}, \mathbf{u}) = \sum_{m=1}^{|B_\alpha|} p(\mathbf{I}_{B,i}^+|\mathbf{I}_{B_\alpha,m}^+)p(\mathbf{I}_{B_\alpha,m}^+|\mathbf{I}_{B,j}, \mathbf{u}) \qquad (5.16)$$

The complete algorithm is shown in Algorithm 10. It makes use of Algorithms 11 and 12, which generate distributions over I-states in $\mathcal{I}_\alpha$ and $\mathcal{I}$ respectively, using sampling as shown in the previous chapter. Algorithms 11 and 12 are trivial modifications of Algorithm 9. We use N_α to denote the number of observation samples used to calculate T_α and N_β to denote the number used to calculate T_β.

Note that, in contrast to Algorithm 5, Algorithm 10 contains two independent loops. The first (steps 1-9) calculates the effects of observation $\mathbf{z}_\beta^+$, while the second (steps 10-24) adds the effects of the action and observation $\mathbf{z}_\alpha^+$. Note also that although Algorithm 10 references T_α, it is not necessary to store the entire CPT because each element is used only immediately after being calculated.

Algorithm 10 Converting a parametric POMDP to a discrete I-state MDP. This Algorithm optimises previous algorithms by pre-calculating the effects of adding information from observation $\mathbf{z}_\beta^+$. It makes use of Algorithms 11 and 12 to generate distributions over posteriors.

1   $T_\beta(\mathbf{I}_{B_\alpha}^+, \mathbf{I}_B^+) \leftarrow 0,\ \forall \mathbf{I}_{B_\alpha}^+ \in B_\alpha,\ \forall \mathbf{I}_B^+ \in B$
2   $foreach\ i \in 1 \ldots |B_\alpha|$
3         $\Delta \leftarrow$ **generateDistributionOverPosteriors**$_\beta(\mathbf{I}_{B_\alpha,i}^+)$
4         $foreach\ l \in 1 \ldots |\Delta|$
5               $foreach\ j \in 1 \ldots |B|$
6                     $T_\beta(\mathbf{I}_{B_\alpha,i}^+, \mathbf{I}_{B,j}^+) \leftarrow T_\beta(\mathbf{I}_{B_\alpha,i}^+, \mathbf{I}_{B,j}^+) + \frac{1}{|\Delta|}\lambda_B(\Delta_l, j)$
7               $end\ foreach\ j$
8         $end\ foreach\ l$
9   $end\ foreach\ i$
10 $T_\alpha(\mathbf{I}_B, \mathbf{u}, \mathbf{I}_B^+) \leftarrow 0,\ \forall \mathbf{I}_B \in B,\ \forall \mathbf{u} \in U,\ \forall \mathbf{I}_{B_\alpha}^+ \in B_\alpha$
11 $T(\mathbf{I}_B, \mathbf{u}, \mathbf{I}_B^+) \leftarrow 0,\ \forall \mathbf{I}_B \in B,\ \forall \mathbf{u} \in U,\ \forall \mathbf{I}_B^+ \in B$
12 $foreach\ i \in 1 \ldots |B|$
13        $foreach\ \mathbf{u} \in U$
14              $\Delta \leftarrow$ **generateDistributionOverPosteriors**$_\alpha(\mathbf{I}_{B,i}, \mathbf{u})$
15              $foreach\ l \in 1 \ldots |\Delta|$
16                    $foreach\ m \in 1 \ldots |B_\alpha|$
17                          $T_\alpha(\mathbf{I}_{B,i}, \mathbf{u}, \mathbf{I}_{B_\alpha,m}^+) \leftarrow T_\alpha(\mathbf{I}_{B,i}, \mathbf{u}, \mathbf{I}_{B_\alpha,m}^+) +$
                               $\frac{1}{|\Delta|}\lambda_{B_\alpha}(\Delta_l, m)$
18                          $foreach\ j \in 1 \ldots |B|$
19                                $T(\mathbf{I}_{B,i}, \mathbf{u}, \mathbf{I}_{B,j}^+) \leftarrow T(\mathbf{I}_{B,i}, \mathbf{u}, \mathbf{I}_{B,j}^+) +$
                                     $T_\alpha(\mathbf{I}_{B,i}, \mathbf{u}, \mathbf{I}_{B_\alpha,m}^+)T_\beta(\mathbf{I}_{B_\alpha,m}^+, \mathbf{I}_{B,j}^+)$
20                          $end\ foreach\ j$
21                    $end\ foreach\ m$
22              $end\ foreach\ l$
23        $end\ foreach\ \mathbf{u}$
24 $end\ foreach\ i$

5.4.1 Computational Complexity

Ignoring the cost of the weighting function and updating the CPTs, the computational complexity of Algorithm 10 is

$$O\left(|B_\alpha|N_\beta\big(C(\mathbf{u}, \mathbf{z}_\beta^+) + N_\beta C(L_\beta)\big) + |B||U|N_\alpha\big(C(\mathbf{z}_\alpha^+) + N_\alpha C(L_\alpha)\big)\right) \qquad (5.17)$$

where $C(\mathbf{u}, \mathbf{z}_\beta^+)$ and $C(\mathbf{z}_\alpha^+)$ are the costs of calculating expected observations in Algorithms 11 and 12 respectively, and $C(L_\beta)$ and $C(L_\alpha)$ are the costs of evaluating the likelihood function for the two observation components. The first term is the complexity of building T_β in the first nested loop of Algorithm 10, while the second term is the complexity of using T_β to build T in the second nested loop. Under the assumption that the cost is dominated by the second term, which loops over actions in addition to beliefs and observations, the cost can be approximated

---

Algorithm 11 generateDistributionOverPosteriors$_\beta$($\mathbf{I}_\alpha^+$)

---
1   $\Delta \leftarrow \emptyset$
2   *foreach* $i \in 1 \ldots N_\beta$
3   sample a state $\mathbf{x}_i^+$ from $p(\mathbf{x}^+|\mathbf{I}_\alpha^+)$
4   calculate the expected observation $\hat{\mathbf{z}}_{\beta,i}^+$ from state $\mathbf{x}_i^+$
5   *end foreach* i
6   *foreach* $i \in 1 \ldots N_\beta$
7   *foreach* $j \in i \ldots N_\beta$
8   calculate $\tilde{L}_{ij} = p(\hat{\mathbf{z}}_{\beta,i}^+|\hat{\mathbf{z}}_{\beta,j}^+)$
9   *end foreach* j
10   *end foreach* i
11   copy the upper triangle of $\tilde{L}$ to the lower triangle
12   normalise each row of $\tilde{L}$ to produce W
13   *foreach* $i \in 1 \ldots N_\beta$
14      $\mathbf{Q}_\beta \leftarrow \emptyset$
15   *foreach* $j \in 1 \ldots N_\beta$
16   add $q^j =<\mathbf{x}_j^+, W_{ij} >$ to $\mathbf{Q}_\beta$
17   *end foreach* j
18   calculate the mean μ and covariance Σ of $\mathbf{Q}_\beta$
19   add $\mathbf{I}^+ =< \mu, \Sigma >$ to Δ
20   *end foreach* i

Algorithm 12 generateDistributionOverPosteriors$_\alpha$($\mathbf{I}, \mathbf{u}$)

---
1   $\Delta \leftarrow \emptyset$
2   *foreach* $i \in 1 \ldots N_\alpha$
3   sample a state $\mathbf{x}^i$ from $p(\mathbf{x}|\mathbf{I})$
4   sample a next-state $\mathbf{x}_i^+$ from $p(\mathbf{x}^+|\mathbf{x}^i, \mathbf{u})$
5   calculate the expected observation $\hat{\mathbf{z}}_{\alpha,i}^+$ from state $\mathbf{x}_i^+$, given $\mathbf{u}$
6   *end foreach* i
7   *foreach* $i \in 1 \ldots N_\alpha$
8   *foreach* $j \in i \ldots N_\alpha$
9   calculate $\tilde{L}_{ij} = p(\hat{\mathbf{z}}_{\alpha,i}^+|\hat{\mathbf{z}}_{\alpha,j}^+)$
10   *end foreach* j
11   *end foreach* i
12   copy the upper triangle of $\tilde{L}$ to the lower triangle
13   normalise each row of $\tilde{L}$ to produce W
14   *foreach* $i \in 1 \ldots N_\alpha$
15      $\mathbf{Q}_\alpha \leftarrow \emptyset$
16   *foreach* $j \in 1 \ldots N_\alpha$
17   add $q^j =<\mathbf{x}_j^+, W_{ij} >$ to $\mathbf{Q}_\alpha$
18   *end foreach* j
19   calculate the mean μ and covariance Σ of $\mathbf{Q}_\alpha$
20   add $\mathbf{I}^+ =< \mu, \Sigma >$ to Δ
21   *end foreach* i

---

by

$$O\left(|B||U|N_\alpha\big(C(\mathbf{z}_\alpha^+) + N_\alpha C(L_\alpha)\big)\right) \tag{5.18}$$

Using a similar approach to measuring complexity, the cost of Algorithm 9 was given in Chapter 4 as

$$O\left(|B||U|N\big(C(\mathbf{z}^+) + NC(L)\big)\right) \tag{5.19}$$

where $C(\mathbf{z}^+)$ and $C(L)$ are the costs of calculating expected observations and likelihoods for the entire observation vector. Algorithm 10 is cheaper than Algorithm 9 because $C(\mathbf{z}_\alpha^+) < C(\mathbf{z})$, $C(L_\alpha) < C(L)$, and potentially $N_\alpha < N$. This difference is quantified for the BlockWorld problem in the following sections.

5.5   Experiments

The approach described in this chapter was applied to the BlockWorld problem presented previously. Recall that the agent was equipped with four range sensors and a collision sensor. All five sensors are conditionally independent given the state. While the distribution over collision observations depends on both the state and the action, the range sensors depend on the state alone.

While it would be possible to use the approach described in this chapter to generate a separate CPT for each of the four range sensors, it will be shown that the cost is dominated by the action-dependent component of the observation. Therefore calculating all four together adds relatively little computation but avoids the approximation inherent in separate calculation. The graphical model is shown in Figure 5.6. The algorithm used to calculate T was essentially Algorithm 10, with $f_{\mathbf{I}_C}$ and $f_{\mathbf{I}_R}$ denoting the transition functions which incorporate the collision observation and range observations, respectively.

We compare against both MDP and the particle-based version of the PPOMDP algorithm as described in Chapter 4. All parameters were kept the same where possible, however precalculating the effect of the range observations introduces two differences.

Firstly, a new set of I-states B_C is required. B_C was chosen to be the same regular grid of I-states used for B, though in general B and B_C needn't be identical.

Secondly, rather than having a single parameter to specify the number of samples to use when calculating $f_{\mathbf{I}}$, there are two: the number to use when calculating $f_{\mathbf{I}_C}$ and the number to use when calculating $f_{\mathbf{I}_R}$. The two parameters needn't take the same value, however 50 samples were chosen for both parameters, just as 50 samples were used in Chapter 4 to calculate $f_{\mathbf{I}}$. In principle the number of particles could be reduced. The particles are used to approximate the

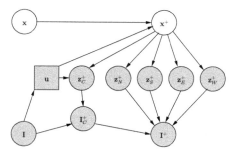

Figure 5.6: Graphical model representing the modified approach to calculating the CPT T governing the Block World. $\mathbf{z}_C^+$ represents the collision sensor, while $\mathbf{z}_R^+ = \{\mathbf{z}_N^+, \mathbf{z}_S^+, \mathbf{z}_E^+, \mathbf{z}_W^+\}$ represents the four range sensors, pointing north, south, east and west respectively.

distribution over possible observations. The approach presented in this chapter splits this into two simpler distributions, each over a lower-dimensional observation space. Since in general the number of samples required to achieve a given density is exponential in the dimension of the space, fewer samples should be needed to approximate each of the simpler distributions to the same accuracy. Given the assumptions used to simplify the computational complexity of Algorithm 10 in Section 5.4.1, the effect of a decrease in the number of samples should be something between a linear and a quadratic decrease in the computational requirements, depending on the relative importance of $C(\mathbf{z}_C^+)$ and $C(L_C)$.

5.6 Results

The results, shown in Figure 5.7, plot both the mean performance per episode and the mean time required to generate a value function, with and without pre-calculation of the effects of the range observations. Observation pre-calculation clearly produces similar or better performance, with a substantial improvement in efficiency (the computational requirements are reduced to approximately one third for this problem). The reason for the improved performance on larger worlds is likely the increased sample density for each individual observation component, as discussed in the previous section.

After factoring the model, the bulk of the computation time was spent calculating the effects of the action and collision sensor. Let T_R denote the CPT describing the transitions resulting from the range observation. The calculation of T_R was found to represent a relatively minor component of the total computational requirements (approximately 10%). This provides support for the complexity assumptions which were made in Section 5.4.1. The remaining 90% was spent calculating T from T_R, which requires a sum over both observations and actions.

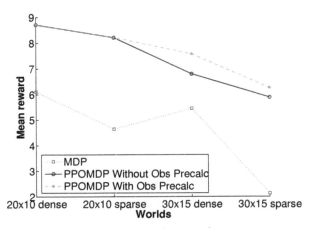

(a) Mean reward per episode

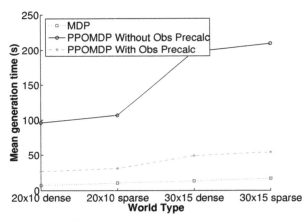

(b) Mean time taken to generate value function

Figure 5.7: Comparison of the PPOMDP algorithm with and without the pre-calculation of the effects of the range observation, showing (a) the mean performance per episode, and (b) the mean time taken to generate the value function, for each world. Both plots average over the four possible goal locations. MDP performance is included for scale.

5.7 Summary

This chapter showed how the conditional independence between observations can be exploited to reduce the total computation required to solve a PPOMDP problem. After formulating the theory, describing the improved algorithm in detail and analysing its computational complexity, the improved algorithm was applied to the BlockWorld problem. The results show similar or better performance, for approximately one third of the computation.

Chapter 6

Using Arbitrary Belief Sets

As discussed in Section 5.1, the approximation made by FVI is to approximate a deterministic transition to an arbitrary belief in a continuous I-space with a probabilistic transition to a discrete set of nearby beliefs. The probability of each transition is set by a weighting function λ which is a core component of the algorithm presented so far.

Until this point, a weighting function based on Freudenthal triangulation has been used, as described in Section 3.3.2. While extremely fast, it requires that the set of beliefs B be a regular grid in parameter-space. This is problematic for scaling to more complicated problems. If the dimensionality of the parameter-space increases (as it will when heading is modelled in Chapter 8), the number of belief points required to cover that parameter-space with a regular grid will increase exponentially. However, as pointed out in Section 1.3, the distribution over beliefs which are likely to be encountered during plan execution is probably not uniform over the entire belief-space. By using an arbitrary set of beliefs rather than a regular grid, the density of belief points can be selected to match this distribution more closely. Scalability then becomes limited by the size of the set of likely beliefs rather than the length of the parameter vector describing each belief.

The impediment to using arbitrary belief sets is that an efficient weighting function is required, and Freudenthal triangulation cannot be used. The key to implementing an efficient weighting function for arbitrary beliefs is to ensure that λ_B operates on only a small neighbourhood of beliefs. If λ_B returns zero weighting for the vast majority of B, that majority need not be considered. Results from the similarity search literature can be used to efficiently find the minority of beliefs for which λ_B returns a non-zero weighting, without exhaustively evaluating λ_B for every element of B.

The remainder of this chapter proceeds as follows. Section 6.1 describes how a weighting function for arbitrary belief sets can be implemented using a kernel function. Section 6.2 reviews data structures from the similarity search literature which can be used to implement this weighting function efficiently. Similarity search in belief-space requires a metric by which to measure the similarity of beliefs. Suitable metrics are discussed in Section 6.3. The efficiency of search using these metrics, and the quality of the resultant plans, are discussed in Sections 6.4 and 6.5 respectively. Section 6.6 presents the results of applying the material above to BlockWorld, demonstrating a significant increase in planning speed. Section 6.7 concludes.

6.1   Efficient Implementation of a Weighting Function

Recall that, as discussed in Section 2.5.2, the following conditions are required of the weighting function λ, for all $\mathbf{I} \in \mathcal{I}$:

$$\sum_{i=1}^{|B|} \lambda_B(\mathbf{I}, i) = 1 \tag{6.1}$$

$$0 \le \lambda_B(\mathbf{I}, i) \le 1 \tag{6.2}$$

We introduce a kernel-based weighting function given by

$$\lambda_B(\mathbf{I}, i) = \frac{1}{C} \eta\big(D(\mathbf{I}, \mathbf{I}_{B,i})\big) \tag{6.3}$$

where η is a kernel function, D is a distance metric returning a scalar distance between two I-states, and the denominator $C = \sum_{i=1}^{|B|} \eta\big(D(\mathbf{I}, \mathbf{I}_{B,i})\big)$ is simply a normaliser to ensure that the conditions in Equations 6.1 and 6.2 are met. A common choice in kernel applications [97] is the Gaussian kernel

$$\eta_g\big(D(\mathbf{I}, \mathbf{I}')\big) = \frac{1}{\sqrt{2\pi\zeta_B^2}} \exp\left(-\frac{1}{2}\frac{D(\mathbf{I}, \mathbf{I}')^2}{\zeta_B^2}\right) \tag{6.4}$$

with an appropriate choice of the bandwidth ζ_B.

This weighting function can be used in the Fitted Value Iteration approximation given by Equation 5.3 and reproduced here:

$$p(\mathbf{I}^+|\mathbf{I}, \mathbf{u}, \mathbf{z}^+) \simeq \sum_{j=1}^{|B|} \lambda_B\big(f_\mathbf{I}(\mathbf{I}, \mathbf{u}, \mathbf{z}^+), j\big)\delta(\mathbf{I}_{B,j}^+) \tag{6.5}$$

Equation 6.5 can be implemented efficiently, without the requirement to examine all of B, if two conditions are met. Firstly, the weighting function needs to return a non-zero weight for only a small minority of B. Secondly, an algorithm is required to find this minority without actually evaluating λ_B for all B.

The first condition is met by choosing a truncated Gaussian kernel

$$\eta\big(D(\mathbf{I},\mathbf{I}')\big) = \begin{cases} \frac{1}{\sqrt{2\pi\zeta_B^2}}\exp\left(-\frac{1}{2}\frac{D(\mathbf{I},\mathbf{I}')^2}{\zeta_B^2}\right) & \text{if } D(\mathbf{I},\mathbf{I}') \le 3\zeta_B \\ 0 & \text{if } D(\mathbf{I},\mathbf{I}') > 3\zeta_B \end{cases}$$

which closely approximates a Gaussian kernel but returns a non-zero value for only those beliefs within a hypersphere of radius $3\zeta_B$. It will be shown how the second condition can be met by using data structures from the similarity search literature to efficiently find the set of beliefs within that hypersphere.

6.2 A Review of Similarity Search Algorithms

Let P be a database of points in the domain V. Assume a query element $q \in V$, and a dissimilarity (or distance) measure D. A common similarity search task is to find the set of elements of P within a hypersphere of some radius r, centred on q. In the context of beliefs, B can be seen as a database. Assuming that a belief is parameterised by the parameter vector $\mathbf{v}$, the domain V is $\Re^{|\mathbf{v}|}$, where $|\mathbf{v}|$ is the length of $\mathbf{v}$. Dissimilarity measures will be discussed in detail in the following sections.

A naive approach to finding the set of neighbours within the radius r is brute-force search: one could calculate the distance from q to every element in P, and remember the set within r. This approach obviously scales linearly with the number of elements in the database.

A great deal of literature exists on the subject of fast, sublinear similarity search (related to the K-nearest-neighbours problem [33]). This section focusses on a small set of the total number of similarity search algorithms which exist, and the particular requirements for fast search in belief-space. For more details, interested readers are directed to one of several survey papers on the subject [26][46][13]. The basic idea for most algorithms is a multi-dimensional generalisation of the idea of binary search. The data is stored in a structure which encodes a partitioning of the domain, allowing regions of the domain to be pruned during search. Section 6.2.1 discusses similarity search in vector spaces, and Section 6.2.2 discusses similarity search in more general metric spaces.

6.2.1 Similarity Search in Vector Spaces: Spatial Access Methods

Spatial Access Methods, or SAMs, provide a means of indexing data items in a multidimensional vector space. They provide efficient access to data, but rely on the assumption that the dissimilarity between objects is based on a distance function which does not include any correlation

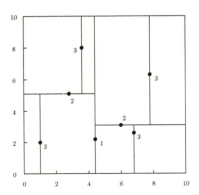

Figure 6.1: A simple kd-tree in $\Re^2$, using Euclidean distance, for 7 points. The lines show how the space is divided into rectangles, the numbering of the points shows the depth in the tree.

(or "cross-talk") between dimensions [38]. More precisely, SAMs assume a Minkowski distance, denoted L_n, examples of which include the Manhattan distance (L_1), Euclidean distance (L_2), and max-norm (L_∞).

Probably the most well-known example of a SAM is the kd-tree [45]. While there are many extensions, such as R-trees [51], Quad-trees [96] and X-trees [9], this discussion is limited to kd-trees since the basic principles are similar.

Briefly, a kd-tree consists of a set of nodes. Each leaf node contains a set of points from P. Each internal node specifies a dimension i and a split value v, dividing the space into two parts with an axis-aligned hyperplane. The left and right children of an internal node are subtrees themselves. The left subtree contains all points in P for which the i'th dimension is less than v, while the right subtree contains all points for which it is greater. The tree therefore defines a partitioning of the space into a set of non-overlapping hyper-rectangles. Figure 6.1 shows a simple two-dimensional example of a kd-tree.

The advantage of using a kd-tree for nearest neighbour search is that parts of the search space can be pruned. As the search proceeds, one can imagine a d-dimensional hyper-sphere of radius τ, centred on q, where τ is the distance from q to the nearest neighbour found so far. If the hyper-rectangle represented by a child node does not intersect this sphere, it can be guaranteed that there is no point in that hyper-rectangle closer than τ. The subtree representing that hyper-rectangle therefore need not be searched. This scenario is illustrated in Figure 6.2.

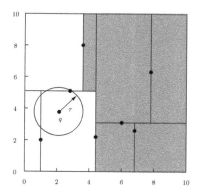

Figure 6.2: The circle bounds the volume of V closer than τ to the query point q, using the L_2 (Euclidean) distance metric in $\Re^2$. Subtrees can be pruned based on the observation that hyper-rectangles which do not intersect this hypersphere can be ignored. This corresponds to the shaded rectangles, which need not be searched.

6.2.2 Similarity Search in Metric Spaces: Metric Indexing Structures

Instead of requiring a vector space, a number of algorithms require only that (V, D) define a metric space. That is, the distance metric D must satisfy the following properties for all $x, y, z \in V$:

- positiveness: $D(x, y) \geq 0$, with $D(x, y) = 0$ if and only if $x = y$.

- symmetry: $D(x, y) = D(y, x)$

- triangle inequality: $D(x, z) \leq D(x, y) + D(y, z)$

These algorithms are commonly referred to as general metric indexing techniques. A disadvantage compared to vector space indexing structures is that they use less information about the data, possibly resulting in poorer performance [13]. Many variants of metric indexing approaches exist [80][116][28][120], and several survey papers have been written on the topic [29][26][55].

In order to discuss metric indexing with reference to a concrete example, this section provides a brief overview of vantage point trees, or simply vp-trees [125]. The issues encountered are relevant for other metric indexing approaches.

A vp-tree is similar to a kd-tree. Each internal node defines a one-dimensional ordering of the database, and splits it in two. Where the kd-tree defines a one-dimensional ordering based on

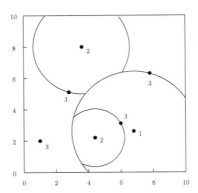

Figure 6.3: A simple vp-tree in $\Re^2$, using Euclidean distance, dividing the same set of points as in Figure 6.1. The circles divide the space, while the numbering of the points shows the depth in the tree. The pivots are the points in the centres of the circles.

the value of a single dimension, the vp-tree orders the data based on the distance from a single point in the database: the vantage point (often known as the pivot). To induce a split in the database, the distances of all points from the vantage point are calculated, and the median is found. Each internal node then stores its vantage point, the median distance, a pointer to the left subtree (containing all points closer to the vantage point than the median distance), and a pointer to the right subtree (containing all points further from the vantage point than the median distance). The resultant partitioning of the domain is shown in Figure 6.3.

While pivots can be chosen at random, a non-trivial increase in search efficiency can be obtained through the use of a good heuristic for pivot selection [125]. Consider the distribution of distances from a candidate pivot to its child points. An effective heuristic is to select pivots which maximise the second moment of this distribution about the median distance.

When performing a nearest-neighbour search, let τ denote the distance from q to the closest point found so far. When searching at a given node, if the distance from q to the median distance is less than τ, then the subtree which does not contain q can be pruned. Another way to put this is that if a hypersphere of radius τ, centred on q, does not intersect the split curve, then half of the search can be pruned. This is illustrated in Figure 6.4. The result is that, at least for a uniformly-distributed dataset in $\Re^2$, search can proceed in $O(\log(n))$ time rather than the $O(n)$ time required for brute-force search.

Unfortunately, the speed-up obtainable by using metric indexing structures decreases with the intrinsic dimensionality of the metric space being searched [26]. Clarkson surveys several methods for estimating the intrinsic dimensionality of a metric space, giving a numerical estimate of the difficulty of searching in that space [29].

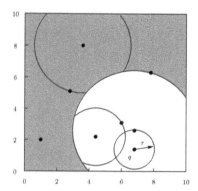

Figure 6.4: The circle bounds the volume of V closer than τ to the query point q, using the L_2 (Euclidean) distance metric in $\Re^2$. Subtrees which do not intersect this circle can be pruned.

6.3 Inter-Gaussian Distance Metrics

The previous section introduced approaches to efficient similarity search, given a distance metric. This section discusses several possible candidates for measuring distances in $\mathcal{I}_{gauss}$. The suitability of each viable candidate is evaluated, both in terms of search efficiency and in terms of the quality of plans which are likely to result from its use.

The evaluation of efficiency highlights the fact that certain distance metrics induce a metric space of high intrinsic dimensionality, and therefore only a very small efficiency increase is possible over naive brute-force search. We present an approach to repairing the metrics in order to avoid this difficulty, and show results which demonstrate that fast similarity search is still possible. The analysis of the expected quality of plans will show that some of the metrics from the literature for measuring distances between general probability distributions exhibit properties which are undesirable for planning in $\mathcal{I}_{gauss}$.

6.3.1 Inter-Gaussian Distance Metrics

This section introduces several commonly-used functions for measuring the dissimilarity between probability distributions, and evaluates their validity for similarity search in $\mathcal{I}_{gauss}$. To help clarify, Figure 6.5 visualises each of the metrics which will be discussed. Each plot in Figure 6.5 shows the distance from various one-dimensional Gaussians to a reference one-dimensional Gaussian with zero mean and unit variance.

Formally, let $p_1(\mathbf{x})$ and $p_2(\mathbf{x})$ denote two probability distributions defined over the continuous domain $\mathbf{X}$, and let $D(p_1(\mathbf{x}), p_2(\mathbf{x}))$ be a function which returns a scalar representing the distance (or dissimilarity) between the two distributions. In order to use a metric indexing technique for fast nearest-neighbour lookups, it is required that $D(p_1(\mathbf{x}), p_2(\mathbf{x}))$ be a true metric, satisfying the three properties given in Section 6.2.2. In addition, since many comparisons will be performed, we require that the distance has a closed-form solution for comparing Gaussians. For the discussion that follows, we assume that $p_1(\mathbf{x})$ and $p_2(\mathbf{x})$ are d-dimensional Gaussians parametrised by the vectors $\mathbf{v}_1$ and $\mathbf{v}_2$, where $\mathbf{v}_1(i)$ denotes the i'th element of $\mathbf{v}_1$. $\mathbf{v}_1$ and $\mathbf{v}_2$ represent the tuples $< \boldsymbol{\mu}_1, \boldsymbol{\Sigma}_1 >$ and $< \boldsymbol{\mu}_2, \boldsymbol{\Sigma}_2 >$, where $\boldsymbol{\mu}$ and $\boldsymbol{\Sigma}$ denote means and covariances respectively.

Parameter-Euclidean Distance

The weighting function used in previous chapters was based on a Freudenthal triangulation. It implicitly assumes a distance metric based on the elements of the parameter vector $\mathbf{v}$ rather than the underlying probability distribution which that vector represents. The extension of this idea to arbitrary belief sets will be called the Parameter-Euclidean distance, denoted D_{PE}, where

$$D_{PE}(\mathbf{v}_1, \mathbf{v}_2) = \left\{ \sum_{i=1}^{|\mathbf{v}|} (\mathbf{v}_1(i) - \mathbf{v}_2(i))^2 \right\}^{1/2} \tag{6.6}$$

D_{PE} is the Euclidean distance, applied in the space of parameter vectors. It is simple, fast to calculate, and defines a vector space, allowing the use of Spatial Access Methods.

KL Divergence

The well-known Kullback-Leibler (KL) divergence is commonly used to measure the distance between distributions. It is given by [30]:

$$D_{KL}(p_1(\mathbf{x}), p_2(\mathbf{x})) = \int_{\mathbf{x}} p_1(\mathbf{x}) \log\left(\frac{p_1(\mathbf{x})}{p_2(\mathbf{x})}\right) d\mathbf{x} \tag{6.7}$$

The KL divergence has the following analytic solution for Gaussians [127]:

$$D_{KL}(p_1(\mathbf{x}), p_2(\mathbf{x})) = \frac{1}{2}(\boldsymbol{\mu}_1 - \boldsymbol{\mu}_2)^T \boldsymbol{\Sigma}_2^{-1}(\boldsymbol{\mu}_1 - \boldsymbol{\mu}_2) + \frac{1}{2}\log\left(\frac{\boldsymbol{\Sigma}_1}{\boldsymbol{\Sigma}_2}\right) + \frac{1}{2}\mathrm{tr}[\boldsymbol{\Sigma}_1\boldsymbol{\Sigma}_2^{-1} - I_d] \tag{6.8}$$

where tr denotes the trace operator, and I_d denotes the d-dimensional identity matrix. Unfortunately the KL divergence satisfies neither the symmetry property nor the triangle inequality. While the KL divergence is often symmetrised by taking the average:

$$D_{KL_{sym}}(p_1(\mathbf{x}), p_2(\mathbf{x})) = \frac{1}{2}[D_{KL}(p_1(\mathbf{x}), p_2(\mathbf{x})) + D_{KL}(p_2(\mathbf{x}), p_1(\mathbf{x}))] \tag{6.9}$$

The failure to satisfy the triangle inequality remains.

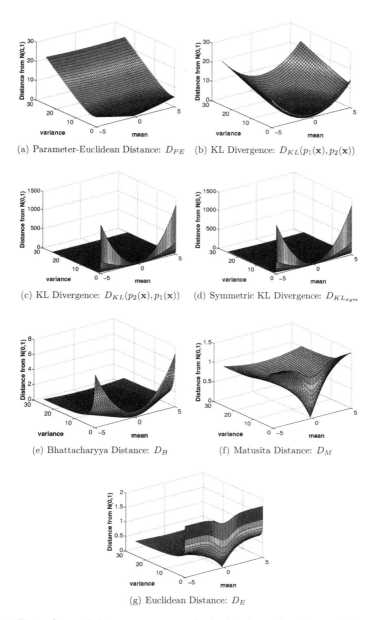

(a) Parameter-Euclidean Distance: D_{PE} (b) KL Divergence: $D_{KL}(p_1(\mathbf{x}), p_2(\mathbf{x}))$

(c) KL Divergence: $D_{KL}(p_2(\mathbf{x}), p_1(\mathbf{x}))$ (d) Symmetric KL Divergence: $D_{KL_{sym}}$

(e) Bhattacharyya Distance: D_B (f) Matusita Distance: D_M

(g) Euclidean Distance: D_E

Figure 6.5: Plots of several distance measures. Each plot shows the distance between various one-dimensional Gaussians (defined by the x-y axes) and a reference one-dimensional Gaussian of zero mean and unit variance. (c) is difficult to see because the height of the surface is dominated by the distances for small σ. Similarly, the distances in (d) are dominated by $D_{KL}(p_2(\mathbf{x}), p_1(\mathbf{x}))$. Only (a), (f) and (g) satisfy the triangle inequality.

α-**Divergence**

The Kullback-Leibler divergence is a special case of the Renyi or α-divergence [54],

$$D_\alpha(p_1(x), p_2(x)) = \frac{1}{\alpha - 1} \log \int_{\mathbf{x}} p_1^\alpha(\mathbf{x}) p_2^{1-\alpha}(\mathbf{x}) d\mathbf{x}, \alpha \neq 1, \alpha > 0 \qquad (6.10)$$

for the special case of α approaching 1:

$$\lim_{\alpha \to 1} (D_\alpha(p_1(\mathbf{x}), p_2(\mathbf{x}))) = \int_{\mathbf{x}} p_1(\mathbf{x}) \log \frac{p_1(\mathbf{x})}{p_2(\mathbf{x})} d\mathbf{x} \qquad (6.11)$$

The α-divergence is essentially a measure of overlap between distributions. Variation of the α parameter produces a continuous range of divergence measures, allowing different features of the distribution to be emphasised.

Bhattacharyya Distance

Another commonly-used special case is the Bhattacharyya distance, related to the α-divergence for $\alpha = 1/2$ [11]:

$$D_B(p_1(\mathbf{x}), p_2(\mathbf{x})) = -\log \int_{\mathbf{x}} \sqrt{p_1(\mathbf{x}) p_2(\mathbf{x})} d\mathbf{x} \qquad (6.12)$$

$$\propto D_{\alpha=0.5}(p_1(\mathbf{x}), p_2(\mathbf{x})) \qquad (6.13)$$

The Bhattacharyya distance is symmetric and has an analytic solution for Gaussians [127]:

$$D_B(p_1(\mathbf{x}), p_2(\mathbf{x})) = \frac{1}{8}(\boldsymbol{\mu}_1 - \boldsymbol{\mu}_2)^T [\frac{1}{2}(\boldsymbol{\Sigma}_1 + \boldsymbol{\Sigma}_2)]^{-1}(\boldsymbol{\mu}_1 + \boldsymbol{\mu}_2) + \frac{1}{2} \log \frac{|\frac{1}{2}(\boldsymbol{\Sigma}_1 + \boldsymbol{\Sigma}_2)|}{|\boldsymbol{\Sigma}_1|^{1/2}|\boldsymbol{\Sigma}_2|^{1/2}} \qquad (6.14)$$

The form of equation 6.14 is intuitively satisfying: the first term represents a quadratic penalty for a difference in means (scaled by the covariance matrices), while the second penalises a difference in covariances. For Gaussians with equal covariances, the second term becomes zero and the distance becomes equal to the Mahalanobis distance (the same is true of the KL divergence) [127]. Unfortunately, the Bhattacharyya distance does not satisfy the triangle inequality.

Matusita Distance

More useful for the task at hand is the Matusita distance [70], also known as the Hellinger distance, given by

$$D_M(p_1(\mathbf{x}), p_2(\mathbf{x})) = \left\{ \int_{\mathbf{x}} \left[\sqrt{p_1(\mathbf{x})} - \sqrt{p_2(\mathbf{x})} \right]^2 d\mathbf{x} \right\}^{1/2} \tag{6.15}$$

It is related to the Bhattacharyya distance by

$$D_M(p_1(\mathbf{x}), p_2(\mathbf{x})) = \left\{ 2\left[1 - \exp(-D_B(p_1(\mathbf{x}), p_2(\mathbf{x}))) \right] \right\}^{1/2} \tag{6.16}$$

and therefore has an analytic solution for Gaussians. Furthermore, it qualifies as a true distance metric.

Euclidean Distance

Any of the L_n distances of the form

$$D_n(p_1(\mathbf{x}), p_2(\mathbf{x})) = \left\{ \int_{\mathbf{x}} |p_1(\mathbf{x}) - p_2(\mathbf{x})|^n \right\}^{1/2} \tag{6.17}$$

satisfy the conditions required for true metrics. This can be seen by observing that the L_n metrics define vector spaces, and a function can be viewed simply as an infinite-dimensional vector. Figure 6.5(g) plots the Euclidean (L_2) distance, which has the following form:

$$D_E(p_1(\mathbf{x}), p_2(\mathbf{x}))d\mathbf{x} = \left\{ \int_{\mathbf{x}} [p_1(\mathbf{x}) - p_2(\mathbf{x})]^2 \right\}^{1/2} \tag{6.18}$$

$$= \left\{ \int_{\mathbf{x}} p_1^2(\mathbf{x})d\mathbf{x} + \int_{\mathbf{x}} p_2^2(\mathbf{x})d\mathbf{x} - 2 \int_{\mathbf{x}} p_1(\mathbf{x})p_2(\mathbf{x})d\mathbf{x} \right\}^{1/2} \tag{6.19}$$

For d-dimensional Gaussians, an analytic solution is available. Each of the first two terms in 6.19 is the integral of the square of a Gaussian, which can be evaluated using

$$\int_{\mathbf{x}} p^2(\mathbf{x})d\mathbf{x} = (2^{2d} \pi^d |\mathbf{\Sigma}|)^{-1/2} \tag{6.20}$$

where $\mathbf{\Sigma}$ is the covariance matrix. The third term is inversely related to the integral of the product of two Gaussians, which can be calculated using:

$$\int_{\mathbf{x}} p_1(\mathbf{x})p_2(\mathbf{x})d\mathbf{x} = \frac{1}{(2\pi)^{\frac{d}{2}} |S|^{\frac{1}{2}}} \exp(-\frac{1}{2}\mathbf{m}^T S \mathbf{m}) \tag{6.21}$$

where $S = \Sigma_1 + \Sigma_2$ and $\mathbf{m} = \boldsymbol{\mu}_1 - \boldsymbol{\mu}_2$.

Note that, while the Euclidean space is a vector space, this does not imply that the Spatial Access Methods of Section 6.2.1 are applicable to the current problem. When using D_E, the space of functions is an (infinite-dimensional) vector space, but the parameter space (defined over $<\boldsymbol{\mu}, \Sigma>$) is not.

6.3.2   The Triangle Inequality

Since several of the metrics discussed fail the triangle inequality, this section attempts to provide some intuition on the subject. Consider the one-dimensional example

$$D_{squared}(x_1, x_2) = (x_1 - x_2)^2 \tag{6.22}$$

as shown in Figure 6.6(a), and the three points $x_a = 0$, $x_b = 1$, and $x_c = 2$. For the triangle inequality to be satisfied, the distance from x_a to x_c via the midpoint x_b should be at least as large as for the direct route. Clearly this is not the case, and therefore the triangle inequality does not hold:

$$D_{squared}(x_a, x_c) = 4 \tag{6.23}$$
$$D_{squared}(x_a, x_b) + D_{squared}(x_b, x_c) = 2 \tag{6.24}$$

Intuitively, the reason is because the journey directly from x_a to x_c encounters a steep slope near $x = 2$, whereas the two smaller journeys encounter only the gentler slope near the origin. The positive second derivative breaks the triangle inequality. In contrast, the constant slope of $D_{linear}(x_1, x_2) = |x_1 - x_2|$ and the decreasing slope of $D_{sqrt}(x_1, x_2) = \sqrt{|x_1 - x_2|}$, shown in Figures 6.6(b) and 6.6(c) respectively, do not break the inequality, even though they preserve the same ordering as $D_{squared}$.

6.3.3   Comparison of Metrics on a Simple Example

This section seeks to provide an intuitive understanding of some important properties of the metrics defined above on a simple example. It will be shown that the properties described here have important implications both for search efficiency and for plan quality.

The example uses four one-dimensional Gaussians, $p_1(\mathbf{x}) = \mathcal{N}(0, 1)$, $p_2(\mathbf{x}) = \mathcal{N}(7.5, 1)$, $p_3(\mathbf{x}) = \mathcal{N}(15, 1)$, and $p_4(\mathbf{x}) = \mathcal{N}(15, 6)$, as shown in Figure 6.7. The distances from $p_1(\mathbf{x})$ to each of the other three Gaussians, as measured by each of the true metrics from Section 6.3.1, are shown in Table 6.1.

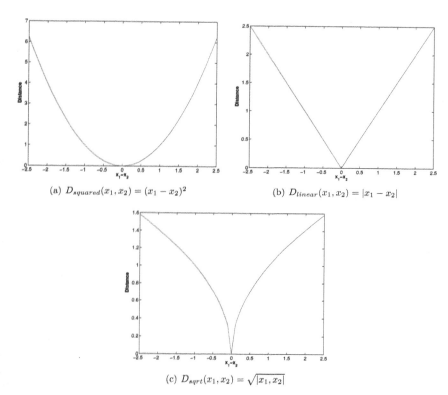

(a) $D_{squared}(x_1, x_2) = (x_1 - x_2)^2$

(b) $D_{linear}(x_1, x_2) = |x_1 - x_2|$

(c) $D_{sqrt}(x_1, x_2) = \sqrt{|x_1, x_2|}$

Figure 6.6: Three one-dimensional distance functions. The triangle inequality holds for only (b) and (c).

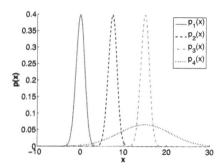

Figure 6.7: The four Gaussians $p_1(\mathbf{x}) = \mathcal{N}(0, 1)$, $p_2(\mathbf{x}) = \mathcal{N}(7.5, 1)$, $p_3(\mathbf{x}) = \mathcal{N}(15, 1)$, and $p_4(\mathbf{x}) = \mathcal{N}(15, 6)$.

	$D(p_1(\mathbf{x}), p_2(\mathbf{x}))$	$D(p_1(\mathbf{x}), p_3(\mathbf{x}))$	$D(p_1(\mathbf{x}), p_4(\mathbf{x}))$
Parameter-Euclidean (D_{PE})	7.500000	15.00000	15.81138
Euclidean (D_E)	0.751125	0.751126	0.630285
Matusita (D_M)	1.413588	1.414214	1.414022

Table 6.1: The distances from $p_1(\mathbf{x})$ to each of the other Gaussians.

Gaussians of Equal Variance

First, consider comparing the Gaussians of equal variance, $p_1(\mathbf{x})$, $p_2(\mathbf{x})$, and $p_3(\mathbf{x})$, using the Euclidean distance metric. Equation 6.19, defining D_E, consists of three terms. For Gaussians of equal variance, the first two terms are constant. The third term (Equation 6.21) is related to the integral of the product of the two Gaussians, giving a measure of the extent to which they overlap. The exponential decay of the Gaussian distribution means that there is virtually no overlap between Gaussians separated by more than 3σ. For non-overlapping Gaussians, the distance becomes dominated by the first two constant terms, meaning that the Euclidean metric does not encode the idea that distributions with more widely-separated means are more dissimilar.

This can be seen from Table 6.1: both $p_2(\mathbf{x})$ and $p_3(\mathbf{x})$ are approximately equidistant from $p_1(\mathbf{x})$ as measured using D_E. In contrast, D_{PE} considers $p_2(\mathbf{x})$ to be much more similar than $p_3(\mathbf{x})$ to $p_1(\mathbf{x})$. This phenomenon is also apparent from Figure 6.5(g). D_E is responsive to small changes in nearby distributions, but approaches a constant value for widely-separated means. The Matusita distance has a similar form, and produces similar results in Table 6.1. Interestingly, it is this behaviour which allows the D_E and D_M to satisfy the triangle inequality, while the Bhattacharyya distance's quadratic penalty for differing means (Equation 6.14) fails the triangle inequality for the reasons outlined earlier in this section. Section 6.4 will show the implications of this behaviour for search efficiency.

Effects of a Change in Variance

Next, consider how distances change as variance is altered. In particular, consider the distance from $p_1(\mathbf{x})$ to each of $p_3(\mathbf{x})$ and $p_4(\mathbf{x})$, which have equal means but different variances. This comparison is shown in the last two columns of Table 6.1. $p_4(\mathbf{x})$ has more overlap with $p_1(\mathbf{x})$, and is therefore considered by D_E to be more similar to $p_1(\mathbf{x})$. In contrast, D_{PE} considers $p_4(\mathbf{x})$ to be less similar to $p_1(\mathbf{x})$ due to the mismatch in variances. Again, the behaviour of D_M is more similar to D_E than to D_{PE}, considering $p_4(\mathbf{x})$ to be more similar to $p_1(\mathbf{x})$. Section 6.5 will discuss how the two approaches to measuring distances affect plan quality.

Gaussian Type	Parameter-Space Dimension	Distance Metric	% Compared
1 Dimensional	2D	Parameter-Euclidean	0.4%
1 Dimensional	2D	Euclidean	1.6%
1 Dimensional	2D	Matusita	1.8%
2D Diagonal	4D	Parameter-Euclidean	1.0%
2D Diagonal	4D	Euclidean	76%
2D Diagonal	4D	Matusita	49%

Table 6.2: Average percentage of the database against which a query element needed to be directly compared when performing a nearest-neighbour lookup using a vp-tree.

6.4 Search Efficiency

Search efficiency using the Euclidean, Matusita and Parameter-Euclidean distances was compared on random databases of size 5000. To generate each point in the databases, each dimension of the mean was sampled from a uniform distribution in the range $[0, 50]$. The diagonal elements of the covariance matrices were sampled uniformly in the range $[0.1, 5.0]$. These ranges were chosen to approximate the kinds of beliefs that might occur for an agent navigating in BlockWorld. For each database, a vp-tree was built. An equivalent number of test points was then generated in the same manner, and the average number of distance calculations required to determine the nearest-neighbour was measured. The results are shown in Table 6.2.

It is clear from Table 6.2 that the efficiency of the vp-tree is significantly worse when comparing distributions than when comparing vectors. Furthermore, the gap between the two increases rapidly with the dimensionality of the parameter vector. When searching for the nearest-neighbours of a two-dimensional axis-aligned Gaussian using D_E, distance calculations must be performed against 76% of the database, which is a disappointingly small improvement over the 100% required for brute-force search.

The cause for this inefficiency can be explained with reference to the simple example from Section 6.3.3. The example showed that for non-overlapping distributions, the Euclidean and Matusita distances do not encode the idea that a wider separation of means implies a greater distance between distributions. Most of the database is therefore considered approximately equidistant. This results in a high intrinsic dimensionality (and hence low search efficiency) for the spaces defined by these metrics.

An Analysis of Intrinsic Dimensionality

Figure 6.8 shows, for each of the scenarios from Table 6.2, the mean histogram of distances from a randomly-selected vantage point to all other points in the database. This histogram is

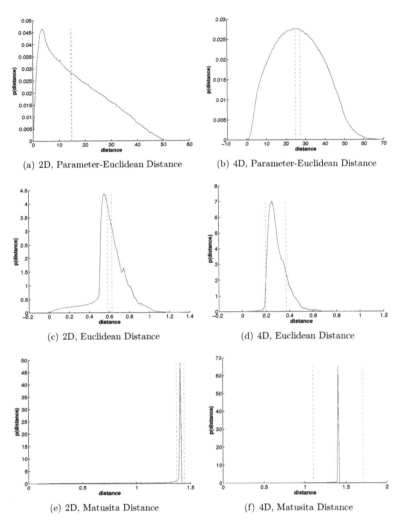

(a) 2D, Parameter-Euclidean Distance   (b) 4D, Parameter-Euclidean Distance

(c) 2D, Euclidean Distance   (d) 4D, Euclidean Distance

(e) 2D, Matusita Distance   (f) 4D, Matusita Distance

Figure 6.8: Each histogram corresponds to a scenario from Table 6.2. The dimensionality refers to the parameter-space, rather than the state-space. Each histogram shows the expected distribution of distances from a randomly-selected vantage point. The dashed lines represent $D = m \pm \bar{\tau}$, where $\bar{\tau}$ is the average distance from a randomly-selected point to its closest point. The databases are of size 5000, with means sampled uniformly from the range $[0, 50]$ and diagonal covariance entries from the range $[0.1, 5.0]$. Note that the shape of these histograms depends on the ranges from which the means and covariances are sampled, and that the expected distance to the closest point depends on the density of points and therefore the size of the database.

a fundamental measure of the intrinsic dimensionality of a metric space [26]. Assuming that the tree was built using random pivots, the histogram gives the distribution of distances from pivots to a random query point q. Let $\bar{\tau}$ denote the average distance from a randomly-selected point to its closest point, and hence the expected distance from q to the closest point in the database. If a hypersphere of radius $\bar{\tau}$ does not intersect with a pivot's split curve, then half of the database can be rejected during a search. Figure 6.8 therefore also shows the band defined by $D = m \pm \bar{\tau}$. Pruning can occur only if q falls outside this band. Therefore the efficiency of searching the vp-tree decreases monotonically with the amount of probability mass covered by this band.

To clarify, one can imagine the extreme case where the distance function returns zero if $p_1(\mathbf{x}) = p_2(\mathbf{x})$, and one otherwise [26]. Under these circumstances the histogram is a single delta at distance 1, entirely covered by $m \pm \bar{\tau}$, and brute-force search cannot be improved upon.

Figure 6.8 explains the observed inefficiency. Considering all non-overlapping Gaussians to be approximately equidistant induces a highly peaked histogram and hence a high intrinsic dimensionality. While this intrinsic dimensionality is low when using the Parameter-Euclidean metric or working with one-dimensional Gaussians, the situation is hopeless when using either the Euclidean or Matusita distance with two-dimensional diagonal Gaussians.

6.4.1 Improving Efficiency Through Metric Repair

The previous section showed that several metrics from the literature induce spaces of high intrinsic dimensionality, which results in low search efficiency. This section suggests an approach to repairing these metrics in order to avoid this problem. It relies on the fact that the sum of two distance metrics is also a distance metric. This is trivial to show: if the three properties of distance metrics from Section 6.2.2 are satisfied by D_1 and D_2, then clearly they are also satisfied by $D_3 = D_1 + D_2$.

Given this fact, the proposed approach is to devise additional metrics which can be added to either the Euclidean or Matusita metrics. In particular, it would be beneficial to devise a metric which heavily penalises widely-separated means or widely-differing covariance matrices. This would ensure that most of the database is not approximately equidistant from any given point, broadening the histograms from Figure 6.8. Additional metrics are chosen such that the original metrics dominate for similar distributions, but the additional metrics dominate for dissimilar distributions. Therefore the ordering for nearby neighbours will remain relatively intact, while the relative distances of the rest of the database can be chosen according to practical considerations.

The Matusita distance D_M was chosen for the subject of this discussion. For the test database of Gaussians described in Section 6.4, it can be seen from Figure 6.5(f) that for Gaussians with similar means, D_M is already relatively sensitive to the differences in covariance that are likely to be encountered. Therefore D_M can be repaired by adding one extra term, D_μ, which penalises widely-separated means:

$$D_{M_R} = D_M + D_\mu \tag{6.25}$$

As a simple distance sensitive to a difference in means, a weighted Euclidean distance between mean vectors was chosen:

$$D_\mu(\mathbf{v}_1, \mathbf{v}_2) = w_\mu \left\{ (meanv_1 - \boldsymbol{\mu}_2)^T(\boldsymbol{\mu}_1 - \boldsymbol{\mu}_2) \right\}^{1/2} \tag{6.26}$$

where w_μ sets the weighting of D_μ relative to D_M. D_μ clearly satisfies all the conditions for a proper distance metric except the requirement that $D_\mu = 0$ if and only if $\mathbf{v}_1 = \mathbf{v}_2$. This condition is not satisfied for vectors with identical means but different covariances, for which $D_\mu = 0$ but $\mathbf{v}_1 \neq \mathbf{v}_2$. It will be satisfied for D_{M_R} however. While the derivation of w_μ is deferred to Appendix A, the final value used was

$$w_\mu = Ms\sqrt{2}\left\{-8\log(1 - s^2)\right\}^{-1/2} \tag{6.27}$$

where $s = 0.9$ and $M = 0.25$.

Experimental Evaluation

The effect of the choice of metric on efficiency was evaluated using a database of the same size and distribution as the database from Section 6.4. 5000 query points were randomly selected from the same distribution. For a set of metrics, the nearest-neighbour of each query point was calculated and the average number of required distance calculations was recorded. The lists of nearest neighbours were then compared, and the percentage of queries on which they agreed was calculated.

The results, shown in Table 6.3, demonstrate that the repaired Matusita metric mostly agrees with the original Matusita metric but requires a much smaller number of distance calculations, comparable with the number required for a general vector space of equivalent dimension. The agreement between D_{M_R} and D_{PE} is not so strong, for the reasons outlined in Section 6.3.3. Note that these results are dependent on the size and distribution of the database. Finally, Figure 6.9 shows the distance distribution for the repaired metric and a plot of the 1-dimensional case, for comparison with Figures 6.8 and 6.5 respectively.

Metric	% Nearest-Neighbour Agreement				% Compared
	D_{PE}	D_E	D_M	D_{M_R}	
Parameter-Euclidean (D_{PE})	100%	69.1%	73.8%	67.6%	0.39%
Euclidean (D_E)		100%	93.8%	95.6%	1.74%
Matusita (D_M)			100%	91.2%	0.87%
Repaired Matusita (D_{M_R})				100%	0.42%

(b) 1D Gaussians (2D Parameter Space)

Metric	% Nearest-Neighbour Agreement				% Compared
	D_{PE}	D_E	D_M	D_{M_R}	
Parameter-Euclidean (D_{PE})	100%	42.7%	49.9%	45.3%	1.02%
Euclidean (D_E)		100%	81.7%	83.5%	75.7%
Matusita (D_M)			100%	86.6%	42.5%
Repaired Matusita (D_{M_R})				100%	1.70%

(b) 2D Diagonal Gaussians (4D Parameter Space)

Table 6.3: For (a) 1D and (b) 2D Gaussians, the central four columns are a matrix showing the percentage of random queries about which pairs of metrics agree on the nearest-neighbour. The rightmost column shows the average portion of the database against which each query had to be compared.

6.5 Evaluation of Expected Plan Quality

While the preceding discussion analysed and improved search efficiency, this section analyses the expected quality of plans when using each metric. The choice of distance metric affects the weighting function, which affects the discrete transition probabilities stored in T. This section will show that certain metrics are more appropriate than others for planning in $\mathcal{I}_{gauss}$, by comparing how well the discrete transitions approximate the underlying continuous transitions.

Each discrete transition (for a given action and observation) assumes that the agent begins at one of the beliefs in B and will transition to another belief in B. Where the true continuous transition does not end at a belief in B, in the agent's mind's eye it will probabilistically 'snap' to a nearby belief after the transition (see Figure 6.10). If this snapping process is inaccurate, the agent will have an unrealistic view of the likelihood of posterior beliefs.

6.5.1 Probabilistic Paths through Discrete Belief Sets

Rather than a single action and observation, consider an initial belief and a fixed set of future actions and observations. This gives rise to a deterministic path through the continuous belief-space. Since the PPOMDP agent plans over a set of discrete beliefs, this deterministic future

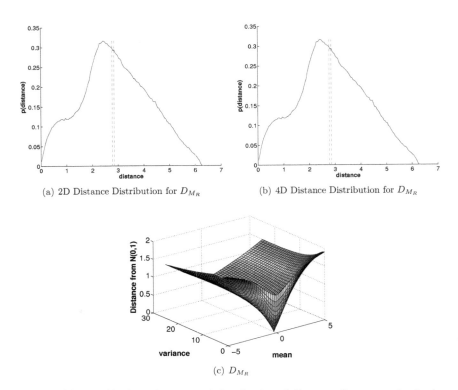

(a) 2D Distance Distribution for D_{M_R} (b) 4D Distance Distribution for D_{M_R}

(c) D_{M_R}

Figure 6.9: (b) and (c) show the expected distribution of distances from a randomly-chosen vantage point. (a) shows the repaired Matusita distance D_{M_R} from a reference 1D Gaussian $N(0,1)$. Compare with Figures 6.8 and 6.5.

path must be approximated by a probabilistic path through B. This is done by approximating each deterministic step in the path by a probabilistic step between discrete beliefs.

A Simple Simulation

This idea can be demonstrated with a simulation, using one-dimensional Gaussian beliefs. The simulation used databases of 5000 and 50000 beliefs, sampled from a uniform distribution. To ensure equal axes on graphs for clarity, the ranges of means and variances were both set to $(0.1, 50)$. Note that a database of 5000 beliefs over this range has one tenth the belief density of the database used in Section 6.4, while a database of size 50000 has equal density. The starting belief was set to $(\mu, \sigma^2) = (35, 20)$. A circular path was then generated, consisting of 50 steps through the continuous parameter-space. This deterministic path is assumed to be the result

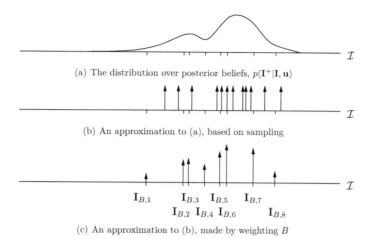

(a) The distribution over posterior beliefs, $p(\mathbf{I}^+|\mathbf{I}, \mathbf{u})$

(b) An approximation to (a), based on sampling

$\mathbf{I}_{B,1}$ $\qquad$ $\mathbf{I}_{B,3}$ $\mathbf{I}_{B,5}$ $\mathbf{I}_{B,7}$

$\mathbf{I}_{B,2}$ $\mathbf{I}_{B,4}$ $\mathbf{I}_{B,6}$ $\qquad$ $\mathbf{I}_{B,8}$

(c) An approximation to (b), made by weighting B

Figure 6.10: An example of a hypothetical one-dimensional belief-space. The figure is identical to Figure 5.2, except that B is non-uniform. (a) shows the true distribution over posterior beliefs, (b) shows a set of sampled posteriors, and (c) shows the approximation to (a) made by FVI, using (b). The true state of affairs is that the appearance of the observation $\mathbf{z}^+$ will result in a posterior belief $\mathbf{I}^+$ being selected from the continuous distribution shown in (a). When planning using FVI, the agent supposes that when the observation is revealed, the posterior belief will be selected from the distribution shown in (c). In other words, the agent is under the impression that its posterior belief will probabilistically 'snap' to one of the beliefs in B.

of a particular set of future actions and observations.

Approximations to this true path can then be sampled, by sampling from the probabilistic discrete transitions which the planner uses to approximate the true continuous transitions. Errors between the true path and sampled paths accumulate because the sampled paths snap to discrete beliefs after every transition, introducing noise. This noise introduces uncertainty about future posteriors beliefs. A good weighting function should minimise this uncertainty, giving the planner an accurate picture of the future.

The weighting function relies on a distance metric and a bandwidth. All true distance metrics from previous sections were compared. The bandwidth was selected as follows. Let $\bar{\tau}_D^B$ denote the mean distance from points in B to the nearest other point in B, using the metric D. ζ_B was set to $C\bar{\tau}_D^B$, where C was set to 0.5.

The results of sampling 100 paths, for each metric and database size, are shown in Figure 6.11. Unsurprisingly, the results show that a higher density of belief points results in a better approximation, with more accurate predictions of future beliefs. More interestingly, Figure 6.11 shows that not all metrics are equal. D_{PE} introduces less noise than the other metrics, which

	5000 Beliefs		50000 Beliefs	
Metric	1D	2D Diagonal	1D	2D Diagonal
Parameter-Euclidean	50.16%	50.29%	49.77%	50.36%
Euclidean	52.87%	67.39%	50.82%	62.08%
Matusita	51.75%	60.14%	50.58%	56.54%
Repaired Matusita	51.38%	59.14%	50.39%	56.13%

Table 6.4: The proportion of trials in which snapping from a random belief increased uncertainty, using databases of size 5000 and 50,000, and using one-dimensional Gaussians and two-dimensional diagonal Gaussians. Each figure is generated from 50,000 trials. An unbiased weighting function will increase uncertainty 50% of the time.

tend to over-estimate the probability of transitions to high-variance beliefs.

The Tendency to Transition to High-Variance Beliefs

D_E, D_M, and D_{M_R} over-estimate the probability of transitions to beliefs with larger variances because they penalise a difference in means more heavily when a potential neighbour has a smaller variance. In other words, when the mean of the belief at the end of a true transition is not aligned with the mean of any belief in B, these metrics will place more weight on the more uncertain nearby beliefs in B. This makes sense for metrics based on overlap. When means are not aligned, more overlap can be attained by selecting a neighbour with a higher variance. Since the Parameter-Euclidean metric is not based on overlap, it does not exhibit this bias.

The bias was quantified more precisely by repeatedly selecting a random point from the same distribution as the database, probabilistically snapping it to a nearby belief in B, and noting whether the snap was to a more or less certain belief. The uncertainty of a belief was measured by calculating the determinant of its covariance matrix, which is monotonically related to its entropy [30]. Table 6.4 shows the results of 50,000 trials. As expected, D_{PE} is the only metric which does not exhibit a bias. The bias for the other metrics is negligible when using a large database in a low-dimensional space, but increases as the density of beliefs decreases (due to either fewer beliefs or the use of a higher-dimensional belief-space).

6.6   BlockWorld Experiments

In this section, the PPOMDP algorithm presented in the previous chapter is extended to allow the use of arbitrary belief sets, and applied to the BlockWorld problem. Due to the factoring described in the previous chapter, the algorithm uses two potentially different sets of beliefs: the set of beliefs after acting and making complete observations, B, and the set of beliefs after acting and observing the output of the collision sensor, B_C. There are therefore potentially

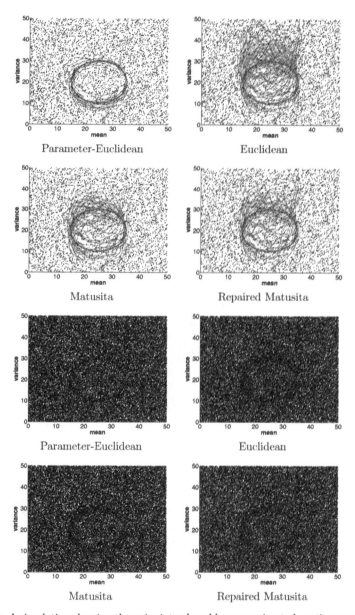

Figure 6.11: A simulation showing the noise introduced by approximated continuous transitions by discrete transitions. (a)-(d) use a database of size 5000, while (e)-(h) use a database of size 50000. Each plot marks the true path through belief-space with a thick blue line, beginning on the right and travelling in a counter-clockwise direction. The sampled approximations are shown as red lines, and the belief set is shown with black dots.

two different weighting functions, λ_B and λ_{B_C}, with two associated bandwidths, ζ_B and ζ_{B_C}. The full list of items which must be specified in order to extend the algorithm by the use of arbitrary beliefs is:

1. the similarity search algorithm;

2. the distance metric D;

3. the kernel bandwidths ζ_B and ζ_{B_C}; and

4. the sets B and B_C.

This section describes experiments with different combinations of settings for these items. The results are presented by first providing details of the combination which was found, empirically, to produce the best results. This consisted of the following:

- **Similarity Search**

  A vantage point tree was used for similarity search.

- **Distance Metric**

  The Euclidean distance in parameter space, D_{PE}, was found to produce the best results.

- **The Bandwidths ζ_B and ζ_{B_C}**

  The bandwidths were set as in Section 6.5. ζ_B was set to $C\bar{\tau}_D^B$, where $\bar{\tau}_D^B$ denotes the mean distance from points in B to the nearest other point in B using the metric D, and C was set to 0.5. ζ_{B_C} was set similarly based on the distances between points in B_C. Performance was found to be relatively insensitive to changes in C.

- **The Sets B and B_C**

  B was calculated by random exploration. The size of the set, $|B|$, was decided *a priori*. Until $|B|$ was reached, a number of episodes were simulated. For each episode, the agent begins at a random unknown valid state. Its belief is initialised by sampling a Gaussian belief, with unit covariance, consistent with the initial state. Each episode lasted 20 iterations. On each iteration, the agent randomly selects an action from a uniform distribution over the continuous range of actions. After taking the action and receiving an observation, it updates its belief. Each new belief is inserted into B. The best results were obtained by choosing B_C to be identical to B. An example belief set is shown in Figure 6.12.

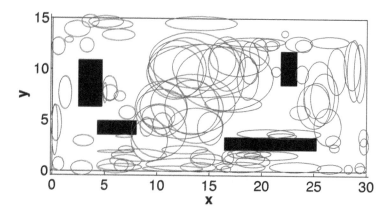

Figure 6.12: A sample of beliefs generated by random exploration. The plot shows 100 beliefs sampled from a belief set of size $|B| = 6000$, using the 30×15 sparse world. For clarity, ellipses bound a one-sigma confidence interval. Generating beliefs through exploration takes advantage of the structure of the problem. For example, only uncertain beliefs are possible in the central open area.

Section 6.6.1 compares the results using this configuration against results from previous chapters. Subsequent sections then describe the effects of deviations from this configuration: Sections 6.6.2, 6.6.3, and 6.6.4 modify the distance metric, set B_C, and belief set generation algorithm, respectively. All results are averaged over the four goal locations specified in previous chapters. The result for each goal location was the average of 1000 episodes.

6.6.1 Comparison Against Previously-Presented Algorithms

Figures 6.13 and 6.14 show the effects of using an arbitrary rather than a regular belief set. Algorithm 10, as presented in Chapter 5, was used to solve the PPOMDP; the only difference is the set B and the associated weighting function λ_B. The parameters for B and λ_B are as specified in Section 6.6, for a range of values of $|B|$.

It can be seen from Figures 6.13 and 6.14 that a reasonable choice for the size of the arbitrary belief set is $|B| = 2000$. For this set size, switching from a regular to an arbitrary belief set makes little difference in terms of attainable reward, perhaps leading to a small improvement for some worlds. The difference in the time taken to generate the value function, however, is considerable. For all four worlds, the reduction in computation time is approximately equivalent to the reduction in the size of the belief set. The regular grids used 8316 and 17856 for the small and large worlds, respectively. Using only 2000 belief points represents a reduction to 24% and 11% respectively, which approximates the reduction in required computation time.

The memory requirements are also significantly reduced. The memory required to store the transition function T for a regular belief set are considerable, even after taking advantage of its sparse nature. For the 20×10 dense world, for each action, the transition function for each belief point referenced an average of approximately 85 next-belief-points. For 33 possible actions, the memory required for T is $8316 \times 33 \times 85 \times 4$ bytes (just under 90Mb) when the index of each next-belief is stored as a four-byte integer. While this could be reduced by ignoring some of the extremely unlikely transitions, it is still considerable and limits the ability of the algorithm to scale to larger problems. In contrast, the arbitrary belief set for the same world referenced an average of only around 14 next-beliefs per belief-action combination. The memory requirements for T are therefore only $2000 \times 33 \times 14 \times 4$ bytes, or about 3.5Mb. In addition, an arbitrary belief set incurs the extra cost of having to store the set B explicitly. This is relatively small however: storing each belief as four floating-point numbers (two for the mean and two for the covariance) consumes just under 8Kb.

6.6.2 Experiments with the Distance Metric

For all four worlds, mean performance and value function generation times were compared for four algorithms: (1) MDP, (2) the version of the PPOMDP algorithm described in Chapter 5 (using a regular grid of beliefs), (3) the same PPOMDP algorithm, using an arbitrary grid of beliefs and the Euclidean distance in parameter space D_{PE}, and (4) the same PPOMDP algorithm, using an arbitrary grid of beliefs and the repaired Matusita distance D_{MR}. Figures 6.15 and 6.16 show, for each world, the results obtained by sweeping across belief sets of many different sizes. Figure 6.17 shows a different view of the same data, obtained by directly comparing performance on each world for $|B| = 2000$.

D_{MR} requires more time for planning than D_{PE} due to the increased computation involved in each distance calculation. Comparing mean reward, D_{PE} out-performs D_{M_R} on most worlds, especially in the 30×15 sparse world. This result is due to the tendency of D_{M_R} to over-estimate the probability of transitions to high-variance beliefs, as was described previously in Section 6.5. This tendency is most apparent in the 30×15 sparse world, where belief density is lowest. Given these results, D_{PE} is adopted as a distance metric for the remainder of this book.

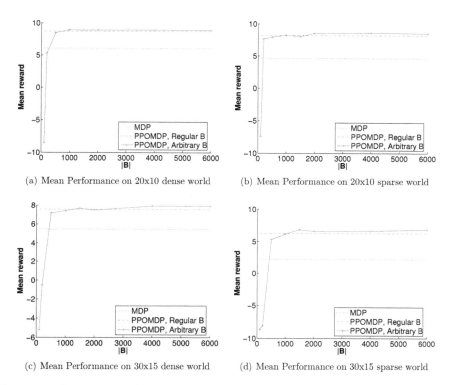

(a) Mean Performance on 20x10 dense world   (b) Mean Performance on 20x10 sparse world

(c) Mean Performance on 30x15 dense world   (d) Mean Performance on 30x15 sparse world

Figure 6.13: A comparison of rewards, for regular and arbitrary belief sets, for various settings of $|B|$. The parameter settings for the arbitrary grid are as described in Section 6.6. The regular grid is as described in previous chapters.

6.6.3 Experiments with a Distinct Set B_C

This section compares the results of two algorithms for generating the sets B and B_C using random exploration. The first is as described in Section 6.6: an agent randomly explores the environment. Each new action and complete observation results in a new belief $\mathbf{I}^+$ which is added to the set B. After exploration is finished, a copy of B is used as the set B_C.

The second algorithm uses distinct sets B and B_C. At each time step, the agent generates two beliefs, $\mathbf{I}_C^+$ and $\mathbf{I}^+$. $\mathbf{I}_C^+$ is the result of the inclusion of information from the action and collision sensor, and is inserted into the set B_C. $\mathbf{I}^+$ is the result of the inclusion of all information, and is inserted into B. Note that $\mathbf{I}^+$ is not actually generated from $\mathbf{I}_C^+$. To do so would incur the approximation penalty of mapping from particles to a Gaussian and back. Instead, $\mathbf{I}^+$ is

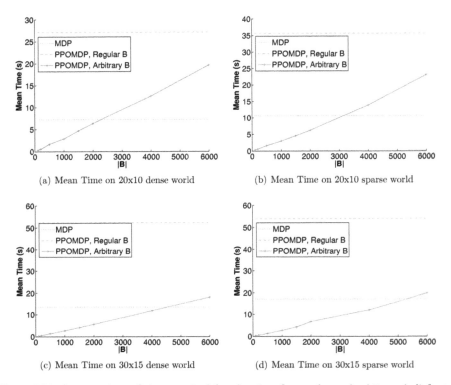

Figure 6.14: A comparison of time required for planning, for regular and arbitrary belief sets, for various settings of $|B|$. The parameter settings for the arbitrary grid are as described in Section 6.6.

calculated directly from $\mathbf{I}$, $\mathbf{u}$ and $\mathbf{z}^+$. The sizes of both sets were fixed at 2000 belief points.

The results are shown in Figure 6.18. The use of distinct set makes little difference to mean reward, but results in a slight increase in the time requirements.

6.6.4   Experiments with Belief Set Generation

It is possible that the belief sets generated by random exploration will be unnecessarily dense in some areas of the belief space, and too sparse in others. This section describes the results of a modification to the belief set generation algorithm to try to prevent this occurrence. The modification is simple: beliefs are added to B only if they are sufficiently different from the beliefs already in B. Specifically, they are added only if the distance to the nearest point in

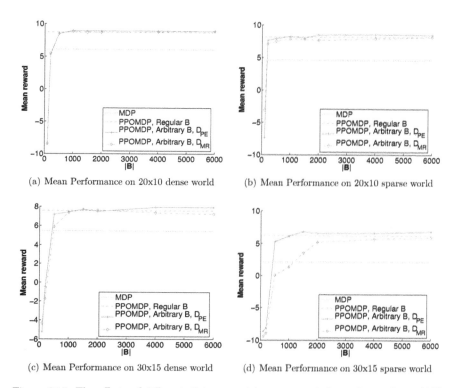

(a) Mean Performance on 20x10 dense world

(b) Mean Performance on 20x10 sparse world

(c) Mean Performance on 30x15 dense world

(d) Mean Performance on 30x15 sparse world

Figure 6.15: The effects of different distance metrics on reward, for various values of $|B|$.

B is greater than a threshold. A similar approach to adding new belief points was used by Thrun [108]. The value of this threshold is a free parameter, for which 0.2 was selected in this work.

Note that this modification to the algorithm for generating the belief set affects the mean distance between beliefs in the set. The algorithm described in Section 6.6 sets the bandwidth ζ_B based on this mean. Therefore comparisons in this section use a fixed bandwidth of 0.5, to ensure that the makeup of the belief set is the only parameter being modified.

The detailed results are shown in Figures 6.19 and 6.20, while a summary for $|B| = 2000$ is shown in Figure 6.21. Figure 6.20 shows that fixing a constant bandwidth has some impact on the time required to generate plans. While previous results showed a well-behaved linear dependence of planning time on $|B|$, Figure 6.20 shows some aberrations.

The results demonstrate that accepting only those beliefs which are sufficiently different from

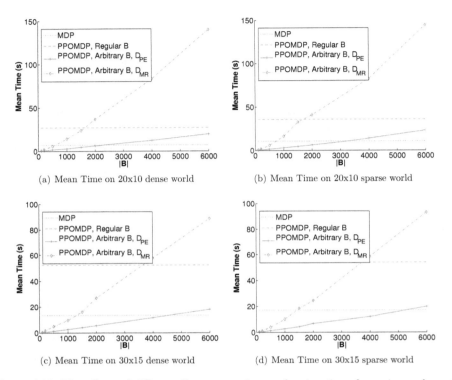

(a) Mean Time on 20x10 dense world

(b) Mean Time on 20x10 sparse world

(c) Mean Time on 30x15 dense world

(d) Mean Time on 30x15 sparse world

Figure 6.16: The effects of different distance metrics on planning time, for various values of $|B|$.

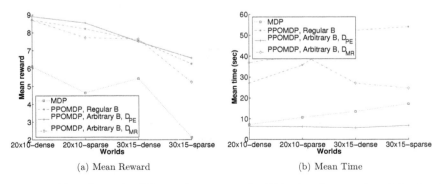

(a) Mean Reward

(b) Mean Time

Figure 6.17: The effects of a different distance metric. This plot shows a slice of the data in Figures 6.15 and 6.16, for $|B| = 2000$.

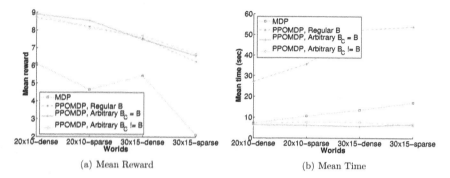

(a) Mean Reward

(b) Mean Time

Figure 6.18: The effects of using distinct sets B_C and B, versus identical sets. Belief set sizes are fixed to $|B| = |B_C| = 2000$. PPOMDP with a regular grid and MDP are included for scale.

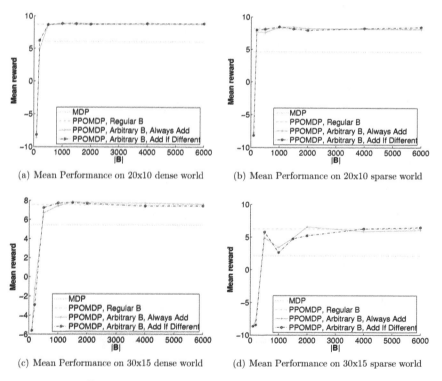

(a) Mean Performance on 20x10 dense world

(b) Mean Performance on 20x10 sparse world

(c) Mean Performance on 30x15 dense world

(d) Mean Performance on 30x15 sparse world

Figure 6.19: The effects of different belief set generation algorithms on reward, for various values of $|B|$.

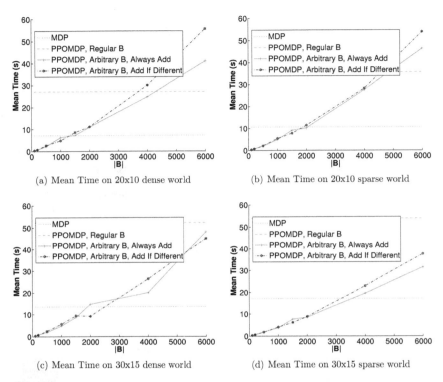

(a) Mean Time on 20x10 dense world

(b) Mean Time on 20x10 sparse world

(c) Mean Time on 30x15 dense world

(d) Mean Time on 30x15 sparse world

Figure 6.20: The effects of belief set generation algorithms on planning time, for various values of $|B|$.

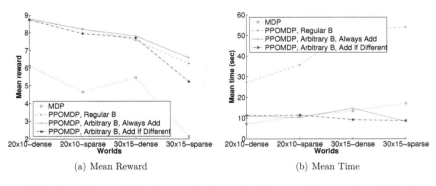

(a) Mean Reward

(b) Mean Time

Figure 6.21: The effects of belief set generation algorithms on (a) reward and (b) time, for $|B| = 2000$. The plots show a slice of the data presented in Figures 6.19 and 6.20.

the existing set does not improve performance, and is detrimental in some cases. In most cases there is a small increase in the time requirements, due to the fact that the exploration algorithm has to execute more steps, since not every step will result in a belief being added to B. This increase could be reduced, since the implementation which generated the results in Figure 6.21 used linear search to find the nearest belief in B. A vp-tree could be used, however it can quickly become unbalanced when points are added incrementally [55]. A better alternative would be to use a data structure optimised for dynamic modification, such as an M-tree [28].

One possible explanation for the disappointing performance is that rejecting overly-similar beliefs represents an attempt to enforce more uniformity on the belief set. The logical extreme of enforcing uniformity is to use a regular grid, for which 2000 belief points are clearly insufficient. Selecting belief points by exploration is advantageous precisely because the set of likely beliefs is in fact non-uniform. By allowing the exploration algorithm to freely choose a set of belief points, the density of B can be matched to the density of beliefs expected during plan execution.

This line of reasoning suggests that one could expect better performance by matching B more closely to the set expected during plan execution. This can be achieved by matching the policy used for exploration to the final policy which will be used for plan execution. While the final policy is not known when selecting B, one could expect to find a better approximation than a random policy. Roy uses two approaches to finding a better approximation: initial exploration using heuristics based on the MDP solution, and further exploration using the POMDP policy after partial calculation of the value function [92]. One must be careful however to ensure that the POMDP planner is given sufficient initial belief points to allow it to make significant deviations from the heuristic plan. This idea will be utilised for the more complex world described in Chapter 8.

6.7 Summary

This chapter presented an approach to efficient function approximation when using arbitrary belief sets. After reviewing the literature on similarity search, it presented several approaches to measuring distances between distributions in general and Gaussians in particular, and discussed their suitability for similarity search in $\mathcal{I}_{gauss}$.

Several of these distance metrics were then evaluated experimentally on a synthetic dataset. Efficiency problems, related to the intrinsic dimensionality when using certain metrics, were identified, analysed and solved. Metrics were then analysed in terms of the expected quality of plans which would result from their use. The metrics based on dissimilarity of general probability distributions were shown to be likely to result in worse plans. This is because they tend to over-estimate the probability of transitions to uncertain beliefs.

After showing how similarity search can be incorporated to implement an efficient weighting function, the approach was experimentally validated on BlockWorld. Individual parameter settings were varied in order to demonstrate their effects.

The results showed a significant speed-up in planning time, plus reduced memory requirements. In addition, there are several less-obvious benefits of using arbitrary belief sets. Firstly, compared to a regular grid, there are fewer free parameters to specify. To specify a set of beliefs, a regular grid requires that the range and level of discretisation of each parameter be specified. An arbitrary grid requires only that the size of the belief set and the algorithm for generating it be specified. Secondly, the results obtained when using a regular grid can be sensitive to the particular details of how grid-points line up with features of interest (such as the goal), which is determined by the range and level of discretisation of the grid. The use of arbitrary grids removes this dependency.

Finally, we note that Pineau has experimented with applying metric indexing schemes to Point-Based Value Iteration, a discrete gradient-based approach [85]. Beliefs over discrete states are stored in metric trees in order to accelerate the comparison of belief points with α-vectors. The results seem to show that the efficiency gains of using metric indexing decrease as the number of states grows, and hence the intrinsic dimensionality of the space increases. This agrees with the results presented in Section 6.3. We expect the use of metric indexing schemes to be more profitable for beliefs described by fewer parameters.

Chapter 7

Plan Execution and Forward Planning

This chapter shows how plan execution can be improved by incorporating online forward planning with prior offline value iteration. Section 3.5 described how the PPOMDP agent's plans have been executed until this point. To briefly review, value iteration requires a maximisation over actions at every belief point. By remembering those maximising actions, an agent executing the plan online can simply apply the maximising action corresponding to the stored belief in B which most closely matches the current belief.

Section 3.5 referred to this strategy as *zero-step lookahead*, referring to the idea that it is a special case of a more general lookahead strategy. Recall that Section 2.7.3 described how a POMDP can be viewed as a game in which turns alternate between the agent selecting an action, and nature selecting an observation. Figure 7.1 shows how this game can be represented as a tree, with circles representing *action-nodes* from which the agent chooses an action, and squares representing *observation-nodes* from which nature chooses an observation. Casting the problem as a game in this way makes it possible to draw on extensive theoretical analysis and results from the AI game-playing literature. The contribution of this chapter is to discuss how results from the game-playing literature can be applied to the PPOMDP formulation, to show how a game-tree can be implemented in an efficient manner for a particle-based PPOMDP, and to experimentally evaluate the approach on BlockWorld.

The remainder of this chapter proceeds as follows. Section 7.1 describes the parallel between POMDPs and game-trees in more detail, and Section 7.2 reviews similar games and strategies from the literature which have proven effective for solving them. Section 7.3 shows how forward planning can be incorporated efficiently into the particle-based PPOMDP formulation presented in previous chapters. Experiments to show the value of incorporating forward planning are described in Section 7.4. The results, presented in Section 7.5, show that online forward planning relaxes the requirements for detailed and time-consuming offline prior planning. Section 7.6 discusses approaches to improving forward planning, and Section 7.7 summarises.

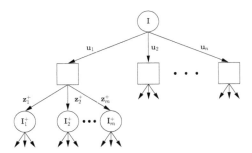

Figure 7.1: A POMDP viewed as a game-tree, starting from belief $\mathbf{I}$. Action-nodes (circles) represent nodes from which the agent chooses a value-maximising action from one of $n = |U|$ choices. Observation-nodes (squares) represent nodes from which the environment probabilistically chooses one of $m = |Z|$ observations. The value of each node is based on the rewards associated with belief-action transitions and the estimated values of the un-expanded leaf nodes.

7.1   Forward Planning as Game-Tree Expansion

A tree represents a plan as follows. Should the agent arrive at a particular action-node, that node dictates both the current action and a future policy:

- At every non-leaf action-node, the current action is the one which leads to the maximum-value observation-node. The future policy is the policy dictated by the child action-node resulting from the current action and the subsequent observation.

- At every leaf action-node, both the current action and all future actions are determined directly from prior value iteration. The maximising action for the nearest belief in B is always selected.

In practice only the first action of any plan will ever be executed, because a new game-tree will be generated for the next action.

In the language of the literature on game-tree searching, the value of a leaf node is determined using an *evaluation function*. This is a somewhat coarse estimate of the value of the node which can be obtained without examining its children, usually through the use of a heuristic. Game-tree terminology also refers to a single level of the tree, consisting of a move by either player, as a *ply*. We prefer to define the depth solely in terms of action-nodes because the evaluation function exists only for action-nodes. Further, we define an *n-step lookahead* plan as a tree of maximum depth n. The plan which has been considered thus far, namely a *zero-step lookahead* plan, corresponds to a tree consisting of a single leaf. The plan is therefore derived solely from the previously-computed value function.

While most games to which game-tree search has been applied use a discrete state-space, the belief-space of the PPOMDP problem is continuous. Plan execution using the value function alone implicitly assumes that the agent begins every action at one of the beliefs in B and will transition to another belief in B. As described in previous chapters, where a transition does not end at a belief in B, in the agent's mind's eye it will probabilistically 'snap' to a nearby belief after the transition (see Figure 6.10). The incorporation of forward planning allows the agent to plan over the entire continuous belief-space for a few moves, delaying this snap. In areas of the belief-space where B is dense, the approximation of snapping to a nearby belief is relatively mild. In sparse areas, however, it is more severe. It will be shown in Section 7.5 how the use of forward planning allows sparse areas of the belief-space to be 'filled in' at execution time.

In game-tree terms, the evaluation function in the PPOMDP case is a function which returns the infinite-horizon discounted cumulative reward when executing the previously-calculated policy while snapping to B at every iteration. Forward planning to a depth of d means that the agent can compare plans which begin snapping to B only after d actions. The scenario is a standard one in planning for game playing: exact forward planning can occur to a certain depth, beyond which a more coarse approximation is used. Problems can occur near the boundary between exact and approximate planning; see the discussion of the horizon problem in Section 7.2.1.

In addition to a continuous state-space, the PPOMDP scenario has a continuous action-space. This has been dealt with by artificially discretising that space. The level of discretisation is a free parameter: a coarse discretisation is preferred in terms of planning time, but may result in poor fine control during plan execution. Forward planning may offer a solution, since the set of actions considered from the root node need not match the actions considered during value iteration. Increasing the level of discretisation during value iteration has a dramatic effect on planning time, because the same number of actions is considered for every belief. However, it may be possible to locally refine the prior plan by considering a larger number of actions, just from the agent's current belief, during plan execution.

As with any tree-search problem, one must choose an order in which to expand nodes when growing the tree, and there are usually better strategies than brute-force breadth-first search. While executing PPOMDP plans for robot navigation, time is a strict constraint: a robot must be able to make decisions quickly in order to interact with the real world, and can therefore choose only a limited number of nodes to expand.

7.2 Strategies for Related Tree-Expansion Problems

Before discussing particular strategies, it is helpful to clarify the nature of the problem and relate it to similar AI problems. The problem of searching the tree resembles many general tree-search problems to which algorithms such as AO* have been applied [78], but differs in several important respects. Firstly, there is no goal state but rather a continuous reward associated with any given path through the tree. Secondly and more importantly, the aim is not to find a path to a goal state, or even a path of maximum reward, but rather to select a good action from the root node. AO* operates by expanding the most promising node first. In contrast, a good tree-search strategy for the problem at hand should begin by expanding nodes which are likely to affect the choice of action from the root. It should try to either lower the value of the apparent best action or increase the values of other competing actions.

In this way the PPOMDP search tree is similar to a minimax tree [95], the differences being that the opponent plays randomly rather than adversarially, and that the state and action spaces are continuous rather than discrete. Given this parallel, it is worth considering effective minimax search strategies. Efficient search of minimax trees has made enormous differences to machines' ability to plan ahead in fully observable deterministic games such as chess [95][118]. It may be worth pointing out once again that the POMDP forward-planning problem, like chess, is fully observable because it is the (observable) I-state which is considered rather than the (unobservable) true underlying state.

In terms of determinism, the POMDP problem is more similar to a game such as backgammon, which has a random element introduced by the dice. Playing against an opponent who plays probabilistically is in fact more difficult computationally than playing against an adversarial opponent as in chess. While an adversarial opponent will deliberately frustrate, he or she is at least predictable. This predictability allows branches of the game-tree to be pruned, since there are many moves a good opponent will clearly never make. A probabilistic opponent can be modelled less precisely, and hence the effective non-prunable branching factor is much higher, limiting the depth of forward search. Comparing successful implementations in different domains, Deep Blue could plan to a depth of around 14 plies in chess [22], whereas TD-Gammon could plan forward only two or three in backgammon [106]. The probabilistic nature of the opponent in the POMDP problem suggests that the horizon of forward planning will be more similar to TD-Gammon's than Deep Blue's.

7.2.1   Common Strategies for MiniMax Tree Expansion

At the heart of many minimax game-playing algorithms is iterative-deepening alpha-beta search [95]. Iterative-deepening avoids the need to specify a maximum depth of search. Specifying a particular depth *a priori* is hazardous because a shallow search will result in a poor plan, but if the search is too deep the agent may run out of time before the search is completed. The results of a completed shallower search are generally more reliable than the results of a partially-completed deeper search [118]. Iterative deepening simply performs a number of fixed-depth searches, beginning with a very shallow search and extending the depth by one ply per iteration.

Alpha-beta search performs a depth-first search of the tree to a specified depth. For each node it maintains bounds on the best outcome that either player can guarantee for themselves for a game passing through that node. Assuming optimal play, a player will never select a branch which is provably worse than the best outcome they can already guarantee. Such a branch can therefore be pruned. Pruning in this manner can result in an effective halving of the branching factor of the tree, although the results are highly dependent on being able to select good actions for evaluation before worse actions [95]. When a probabilistic element is involved, as in backgammon, alpha-beta search can still be used. The only difference is that guarantees are much harder to provide because future play is less predictable [95].

Transposition tables are often used to further limit the branching factor [95]. A transposition table is a lookup table of game positions which a planner can use to avoid repeated searching of identical positions which are reachable via different sequences of moves, or in cycles. While transposition tables can make a large difference for discrete games, it is less clear how to apply them to games in a continuous state-space.

Search to a fixed depth d can suffer from the horizon problem [95]. One manifestation of this problem is a plan which involves significant events at depth d. For example, a chess plan may end with the capture of a supported bishop by the queen. The evaluation function may assign high value to the resultant position, failing to see that the the opponent will immediately capture the queen on the next move. Another manifestation involves an inevitable calamitous event which the evaluation function cannot foresee. If the planner can push the event over the edge of the forward planning horizon, it will apparently disappear. Therefore an apparently-successful strategy is to waste time, or make slightly-detrimental stalling moves which only delay the inevitable.

The horizon problem arises due to the mismatch between the crude evaluation function and accurate forward planning. While the problem can be lessened by an improved evaluation function, it cannot be eradicated without a perfect evaluation function, which would render forward search redundant. A solution to the former manifestation is quiescence search, a specific case of a more general technique known as singular extension [118]. The use of singular extensions involves searching more thoroughly and deeply in areas of game-trees near significant events. These tend to correspond to scenarios in which the evaluation function is less stable. Examples of significant events include domain-specific events such as check or capture in chess. More generally, a node whose value is significantly higher than its siblings' (indicating a forced move) tends to be a good candidate for further search.

Upton overviews a number of more advanced techniques for identifying significant areas of the game-tree on which to focus [118]. He suggests however that once an area has been deemed worthy of special attention, it should be searched thoroughly. The most thorough means is brute-force full-width fixed-depth search. Since the area near the root is certainly significant, a successful strategy seems to be to begin any search with an initial shallow full-width search from the root. For PPOMDP game-trees for robot navigation, considering the high branching factor arising from a probabilistic opponent plus the tight constraints on the amount of time available for online planning, it seems unlikely that enough time will be available to move beyond this initial full-width search.

7.3   Forward Planning for a Particle-Based PPOMDP

As mentioned in Section 2.7.3, the depth of a breadth-first search is limited by the exponential explosion of the number of nodes at each level. Specifically, the number of nodes at depth d is $(mn)^d$, where m and n are the branching factors at observation and action nodes respectively. To maximise the forward planning horizon, it is therefore important to minimise the storage and computational costs at each node. Note that it is assumed for the purposes of this discussion that the reward is action-independent; the extension to action-dependent rewards is straightforward.

7.3.1   Logical Tree Structure

In principal, the structure of the tree is as follows. Every node needs to store:

- the belief at that node, either in parametric form or as a set of samples; and
- the estimated value of the node, denoted $\hat{v}$.

In addition, observation-nodes must store

- a set of arcs to action-nodes, with associated observations and probabilities

and action-nodes must store

- a set of arcs to observation-nodes, with associated actions; and
- the reward associated with the belief at that node, denoted r.

For accuracy reasons, it is preferable to propagate beliefs forward as sets of weighted samples rather than incur the approximation error involved in mapping to and from parametric form.

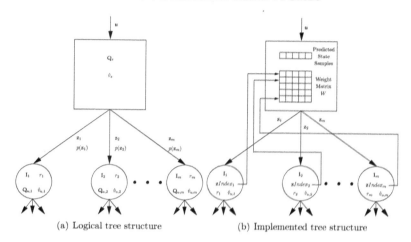

(a) Logical tree structure   (b) Implemented tree structure

Figure 7.2: The structure of a small segment of a particle-based PPOMDP game-tree for forward planning, showing the information stored at each node. (a) shows the logical structure, whereas (b) shows the structure which was implemented. The two differ for efficiency reasons. Squares are observation-nodes, circles are action-nodes. Logically, each observation-node stores a particle set $\mathbf{Q}_z$ plus an estimated value $\hat{v}_z$, and each action-node stores an I-state $\mathbf{I}$, a reward r, a particle set $\mathbf{Q}_u$, and an estimated value $\hat{v}_u$. Each arc stores its action or observation, plus arcs leaving observation-nodes store the probability of that observation.

Therefore every node stores a set of particles $\mathbf{Q}$ where each particle is a tuple $q_i = <\mathbf{x}_i, w_i>$, with w_i specifying the weight.

In addition to storing beliefs as particle sets, action-nodes need an associated parametric belief. This is required to estimate the value of the node from the value function before it is expanded. The structure of the tree, showing the information which is logically associated with each node and arc, is shown in Figure 7.2(a). As will be described in subsequent sections, the actual implementation differs from this logical structure for efficiency reasons.

7.3.2 Observation-Nodes

Since there is no way to estimate the value of an observation-node directly from the prior value function, it must be estimated from its child action-nodes. All leaf observation-nodes must therefore be expanded immediately. The estimated value $\hat{v}_z$ of an expanded observation-node with particle set $\mathbf{Q}_z$ is the weighted sum of the estimated values of its children:

$$\hat{v}_z = \sum_{i=1}^{m} p(\mathbf{z}_i | \mathbf{Q}_z) \hat{v}_{u,i} \tag{7.1}$$

where $\hat{v}_{u,i}$ is the value of the i'th child action-node.

In analogy with the approach to belief propagation presented in Chapter 4, the number of arcs leaving an observation node is set to the number of particles, with each observation being the expected observation of the corresponding particle. The observation branching factor is therefore $m = |\mathbf{Q}_z|$. It will be shown that $\mathbf{Q}_z$ contains uniform sample weights, and hence the term $p(\mathbf{z}_i|\mathbf{Q}_z)$ can be replaced with a constant.

Each observation results in a new weighted particle set in each child action-node. However, since the observations don't alter the state samples but rather adjust the weights, the child action-nodes all share a common set of state samples. In addition, one can take advantage of the symmetry of the likelihood function by pre-calculating a weight matrix W in the parent observation node. For details of the construction of W, the reader is referred to Section 4.4.2.

Given these potential optimisations, the actual structure of the tree is as shown in Figure 7.2(b). The state samples for both child action-nodes and parent observation-nodes are identical, and are therefore stored only once in the observation-node. The weights in the observation-node are uniform, and need not be stored explicitly. The weights for the action-nodes are stored in the parent observation-node's weight matrix. Each action-node contains an index to the row of W that specifies its weights.

Computing the statistics of the state samples, weighted by the appropriate row of W, produces the action-node's parametric belief $\mathbf{I}$. This parametric belief is used to calculate the new leaf action-node's estimated value, $\hat{v}_u$, from the prior value function.

7.3.3   Action-Nodes

When unexpanded, an action-node's value is estimated using

$$\hat{v}_u = \sum_{j=1}^{|B|} \hat{V}(\mathbf{I}_{B,j})\lambda_B(\mathbf{I}, j) \tag{7.2}$$

where $\hat{V}$ is the previously-stored value function. When it has at least one child, its estimated value is a maximisation over its n children:

$$\hat{v}_u = \gamma \max_{i \in n} \hat{v}_{z,i} \tag{7.3}$$

where $\hat{v}_{z,i}$ is the estimated value of the i'th child observation-node, evaluated using Equation 7.1.

An action-node is expanded by selecting an action $\mathbf{u}$, then predicting the particles $\mathbf{Q}_u$ forward according to $\mathbf{u}$ to produce the new particle set $\mathbf{Q}_z$. This is done using Algorithm 13. Since Algorithm 13 samples from $\mathbf{Q}_u$ with probability proportional to the particle weights (in step

Algorithm 13 Generates a particle set for an observation node, $\mathbf{Q}_z$, based on an action-node's particle set $\mathbf{Q}_u$ and an action $\mathbf{u}$. The number of particles representing the observation node's belief is set to $\gamma^2|\mathbf{Q}_u|$. Using the discount factor γ causes nodes at greater depths to use fewer particles.

1 *for* $i \leftarrow 1\dots\gamma^2|\mathbf{Q}_u|$
2   sample a state $\mathbf{x}$ from $\mathbf{Q}_u$, with probability proportional to particle weights
3   sample a predicted state $\mathbf{x}^+$ from $p(\mathbf{x}^+|\mathbf{x}, \mathbf{u})$
4   add the tuple $< \mathbf{x}^+, w >$ to $\mathbf{Q}_z$, where w is a uniform weight
5 *end for*

2), and the prediction step does not alter the weights, the resultant particle set $\mathbf{Q}_z$ has uniform particle weights. Enforcing uniform weights in this prediction step serves the same role as resampling in particle filters, namely combatting degeneracy (insufficient variance in the weights) [4]. When implementing a particle filter, a common strategy is to resample only when the effective sample size drops below a threshold, rather than at every time step. In this case, however, uniform weights are always enforced in order to simplify matters for the observation nodes, as described in Section 7.3.2.

The number of samples remains to be specified. Kearns *et al.* suggest that fewer samples are necessary deeper in the tree, since the impact on the top-most values are diminished due to the discount factor γ [62]. Applied to the current problem, if the direct children of the root action-node use $|\mathbf{Q}_z| = K$ samples, then the children of the action-node at depth d use $\gamma^{2d}K$ samples. This produces the factor γ^2 in step 1 of Algorithm 13.

7.4 Experiments

Experiments were performed on the BlockWorld problem from earlier chapters. Unless otherwise stated, the best set of parameters from earlier chapters was used. Tests were performed under three conditions:

1. considering 33 possible actions during value iteration, with no forward planning during execution;

2. considering only 9 possible actions during value iteration, with no forward planning during execution; and

3. considering only 9 possible actions during value iteration, but planning forward with 33 possible actions during execution.

The first option above is identical to the best algorithm from the previous chapter. The set of actions is extensive and the algorithm performs well. The second option should produce a plan much more quickly, but the quality is likely to suffer. Eight moves of 2m are considered, spread uniformly over the range $[-\pi, \pi)$, plus the move $(0.1m, 0)$. Finally, it is hoped that the third option will provide the best of both worlds, quickly producing a coarse plan then locally filling in the details during plan execution.

The number of samples at the root node was chosen to be 50. The time allowed for online decision making was 10ms. Note that online decision-making times were measured using wall-clock time, whereas offline plan generation times were measured using CPU time. While a longer online decision-making time would be acceptable, the BlockWorld problem is intended as a test-bed for the development of algorithms for eventual application to real-world problems. The figure of 10ms was chosen by selecting an approximate upper limit of 100ms as an acceptable decision-time threshold in more realistic problems, and estimating that action-node expansion in those problems might require an order of magnitude longer.

Measurements of the time taken to consider an action followed by the subsequent observations showed that the full 10ms was required to expand the 33 actions from the root node plus their resultant observation-nodes. Therefore all experiments used a tree of just three levels: the root action-node, followed by 33 observation-nodes, followed by $33 \times 50 = 1650$ action-nodes. As suggested in Section 7.2.1, the search begins with a shallow full-width search from the root. Unfortunately, this strategy leaves no remaining time in which to try more selective search strategies. Performing a fixed-depth search in this way may result in manifestations of the horizon problem. Section 7.6 suggests approaches to combatting this by improving forward-planning speed to allow better search strategies.

7.5 Results

Figures 7.3 and 7.4 show the mean performance and time requirements, respectively, for each of the three variations described in the previous section. Again, the results show an average over the four goal configurations described in Chapter 3, with 1000 episodes being used for each goal configuration.

The most salient point from Figure 7.3 is that similar performance can be attained by considering a larger number of actions during plan execution as by prior planning with that many actions, assuming a reasonable number of belief points. Figure 7.4 shows that the time saved by prior planning using a smaller number of actions is considerable. In other words, online planning confers the benefits of precise offline planning with the consideration of many actions, but does not incur the associated cost. Note that the times shown in Figure 7.4 include the time taken for belief set generation, hence a linear relation to the number of actions is not expected.

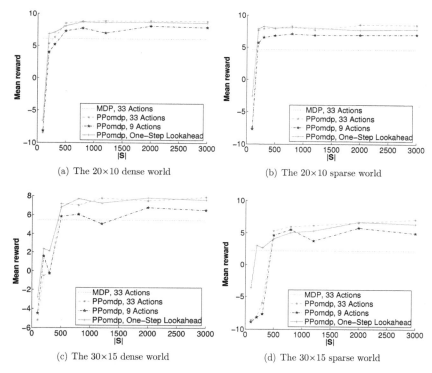

(a) The 20×10 dense world

(b) The 20×10 sparse world

(c) The 30×15 dense world

(d) The 30×15 sparse world

Figure 7.3: For the four worlds and a variety of belief set sizes, a comparison of mean rewards using (i) only 9 actions, (ii) 33 actions, and (iii) only 9 actions for value iteration but all 33 actions for one step of forward planning. The results of using 33 actions for MDP planning are included for scale.

A second point of interest is the behaviour when the belief set is relatively sparse. This is easier to see in Figure 7.5, which shows the mean performance for two belief set sizes, $|B| = 200$ and $|B| = 2000$. As was seen previously, a certain minimum density of belief points is required before the PPOMDP planner begins making sensible plans. These results suggest that when incorporating one step of forward planning, the minimum density requirement is lowered. Figure 7.5(a) shows that, for the 30×15 sparse world, the threshold for sensible plans is greater than 200 beliefs. Without forward planning, PPOMDP produces terrible results. However, a forward-planning agent can still attain a reward similar to the reward attainable by the MDP-based agent. We expect this result to be important as we move to more realistic worlds in Chapter 8, for which B will necessarily be less dense than desired due to the size and dimensionality of the state-space.

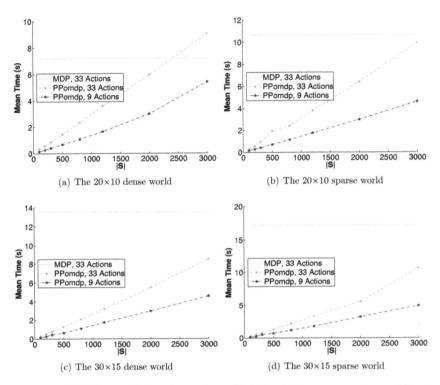

Figure 7.4: For the four worlds and a variety of belief set sizes, a comparison of the times required to produce a plan using (i) only 9 actions, and (ii) 33 actions. The results of using 33 actions for MDP planning are included for scale.

7.6   Planning Further Ahead

A major cause of the computational difficulty of forward planning in this problem is the large branching factor for observation-nodes. Coupled with this is the fact that no evaluation function exists for observation-nodes, necessitating their immediate and complete expansion. The computational cost could be reduced substantially, and hence an agent could plan further ahead, if this branching factor could be reduced. There are two likely approaches to achieving this: using fewer observation samples, and detecting identical observations.

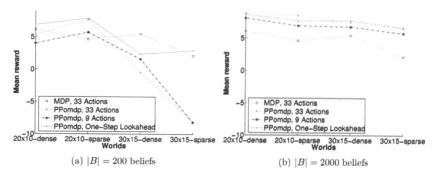

(a) $|B| = 200$ beliefs (b) $|B| = 2000$ beliefs

Figure 7.5: This figure shows a different view of the same data as Figure 7.3, for (a) $|B| = 200$ and (b) $|B| = 2000$ beliefs.

Using Fewer Observation Samples

The implementation described in this chapter required that each observation-node describe the weights of the particles of its children with a square weight matrix. A square matrix was chosen in order to take full advantage of the symmetry of the likelihood function, as described in Chapter 4. Forcing W to be square implies that the number of particles defining the distribution in each child action-node is equal to the number of observation arcs leaving the action-node. When choosing the number of particles to use, one must therefore satisfy two constraints with the same number:

1. there must be enough samples to represent the distribution over observations adequately; and

2. there must be enough samples to represent each belief adequately.

While the appropriate number is problem dependent, it is likely that one of the two constraints will dominate. For the BlockWorld problem, the most pressing constraint appears to be the latter. 50 samples seems like a reasonable number to represent each belief, but seems excessive for representing distributions over observations. Choosing a non-square matrix would allow the above constraints to be satisfied with two different parameters. At the cost of exploiting symmetry less fully, the branching factor could be dramatically reduced. Assuming a non-square W, a possible extension would be to use an adaptive sampling technique such as KLD sampling [42] to set the number of observation samples depending on the complexity of the observation distribution.

Detecting Identical Observations

In general, all observations in the BlockWorld problem will be different, due to the continuous range component. However in areas of the belief-space where range observations are unlikely and observations differ only in the discrete collision sensor, repeated observations will occur. In areas devoid of obstacles, only one observation (no collision and no range observations) is possible. The implementation described above assumed a uniform weighting for each of the arcs leaving an observation-node, requiring that repeated observations be represented explicitly with child action-nodes. Large computational savings would be possible if non-uniform weights were allowed.

While continuous observations will never be identical, it is likely that they will often be similar. Discretising the observation-space with a fine grid introduces a small approximation but may have a significant impact on the branching factor by forcing similar observations to be identical. Note that any extra computation time spent searching for identical observations is likely to pay off due to the geometric increase in the number of nodes at each level in the tree.

7.7 Summary

This chapter drew parallels between POMDPs and sequential games, and discussed game-tree search as a common solution method. It reviewed a number of algorithms for efficient tree search, and discussed their applicability to the POMDP problem. The implementation of game-trees for the particle-based PPOMDP approach was presented and then experimentally evaluated. This showed that the incorporation of forward planning over a short horizon at execution time relaxes the requirements for precise prior offline planning. A coarse plan can be produced quickly, then combined with forward planning to produce results similar to those attainable by a detailed time-consuming offline planning process. Suggestions were then presented for methods of improving the efficiency of forward planning, with the potential to increase the online planning horizon.

The following chapter presents a complex real-world planning problem, which will require all the improvements discussed up to this point. It is anticipated that online forward planning will be particularly beneficial, because prior plans will necessarily be coarse by virtue of the complexity of the problem.

Chapter 8

Scaling to the Real World

The BlockWorld example has served as a simple scenario where algorithms could be tested and compared easily. However, as stated in Chapter 1, the eventual aim is to apply PPOMDP planning to a real robot navigation problem. This chapter achieves that aim incrementally, by adding realistic dynamics to a toy problem, then simulating a real environment, and finally presenting results from running live on a real robot.

To this end, Section 8.1 provides a more detailed description of the application domain which was briefly introduced in Chapter 1. Section 8.2 discusses some of the challenges for navigation in this domain, and outlines the approach taken for localisation. A more precise, mathematical description of the process and sensor models is given in Section 8.3. It shows how the PPOMDP formulation can be extended from BlockWorld to a real-world application.

Section 8.4 explains how policies are evaluated. It describes the framework used for evaluating policies in both realistic simulations and reality, the models used during realistic simulation, and the non-probabilistic policy against which PPOMDP is compared. Section 8.5 presents ToyWorld, a toy problem with realistic dynamics, and Section 8.6 presents RealWorld, the final problem to which PPOMDP will be applied. RealWorld experiments are carried out both in simulation and during live execution on a robot. Results show that the problem is tractable for PPOMDP, and that PPOMDP represents a significant increase in reliability over a non-probabilistic planner. To the author's knowledge, this represents the most challenging robot navigation scenario to which POMDP solution methods have successfully been applied to date. Section 8.7 summarises the chapter.

8.1 Application Domain

The target problem is an industrial application involving several robots navigating in a known mock-up urban environment for long periods of time (episodes on the order of eight hours continuous operation). The task involves having the robots visit a series of waypoints on a specified time schedule, at speeds up to 2 metres per second. The environment and robots are shown in Figure 8.1. The robots are based on the Segway RMP. In addition to their wheel encoders and on-board inertial system used for balancing, each carries a single sensor: a horizontally-mounted forward-pointing SICK laser.

The use of a dynamically-stabilised platform introduces several difficulties, but allows the robots to have a high centre-of-mass for a relatively small base. This allows the robots to be of approximately human height while still fitting through doorways designed for humans. An obvious disadvantage of using a dynamically-stabilising platform is the possibility that robots might fall over, a scenario from which they are unable to recover. While the risk is negligible when traversing flat level terrain, the robot is totally incapable of traversing a step-change of more than a few centimetres. A curb, for example, is completely non-traversable. In addition, the risk of falling is serious when traversing small undulations at high speed. If a wheel leaves the ground for more than a moment, the vehicle will lose the ability to control its pitch and will fall.

Unfortunately, the robots' sensors are unable to detect terrain which might cause a fall. The sensors are mounted high, where they are most useful for localisation. The approach taken to avoiding falls is to engineer the working environment of the robots such that it is hazard-free, and rely on the robots' navigation capabilities to avoid known dangerous areas. This can be non-trivial because a number of non-detectable hazards exist just outside of the robots' designated working area, as shown in Figure 8.2.

Since the application is real, the number of robots is high, and the time-scales are long, reliability is clearly a major concern. In general, failures in the system can be attributed to one of three areas: hardware-related, software-related, or algorithmic. The former two areas are implementation issues, and outside the scope of this book. Algorithmically, most of the reliability challenges revolve around the localisation module. As will be described in Section 8.2, the environment and robots have certain characteristics which make localisation particularly difficult.

Two sources of localisation-related error are likely. The first is that the localisation module itself will fail by becoming inconsistent. That is, if the localisation module considers the true robot pose to be sufficiently unlikely, the filter will have difficulty recovering. The second source of error arises from the fact that the robot has real tasks to perform, and its ability to carry out these tasks depends on its unobservable true state. The default controller, however, is based on the invalid assumption that the estimated maximum-likelihood state is true. When localisation uncertainty is small this assumption is not problematic, but we will show that control based solely on the maximum-likelihood state can result in catastrophic failure when the true state is toward the tails of an uncertain distribution. We will also show how a POMDP-based controller,

(a) two robots, foreground and background

(b) a portion of the environment

Figure 8.1: The robots and a portion of their environment. The vertical white strips are retro-reflective fiducials to aid navigation.

(a) non-traversable curbs   (b) a non-traversable gutter

(c) non-traversable rough terrain

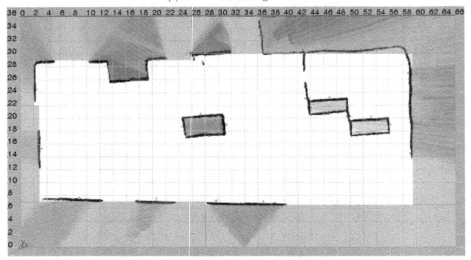

(d) an occupancy grid map of the environment, measuring 66m × 36m

Figure 8.2: On the border of the robots' designated working area lie a number of non-detectable hazards, such as (a) curbs, (b) a gutter, and (c) rough terrain. The former two are completely non-traversable, while the latter is non-traversable at speed. (d) is an occupancy grid map of the environment. The dark yellow regions indicate potentially hazardous areas which the robot should not enter. The small red squares indicate laser-reflective fiducials.

capable of managing its entire probability distribution, can significantly improve the reliability of the system.

8.2 Localisation Approach

For the reasons discussed in Section 4.2, a particle filter is used for localisation. Specifically, we use a KLD particle filter, which dynamically adapts the number of samples [42]. Unlike previous chapters, the filter used for online estimation (as opposed to planning) does not convert to and from a Gaussian representation at each iteration, and hence does not lose the detailed information stored in the particles. A Gaussian is generated in order to make decisions, but the particles are retained. In order to generate a Gaussian for decision-making, the localiser first clusters particles into groups, then calculates the mean, covariance, and total weight of each group. The Gaussian with the greatest weight is used for decision-making. Approaches to using all Gaussians for decision-making, rather than simply the most likely, will be discussed in Chapter 9.

In general, a particle filter requires the specification of a process model and a likelihood function. This section discusses some of the difficulties faced when attempting to localise in this particular environment, and motivates the general approach which was finally adopted. Specific definitions of the process model and likelihood function will be presented in Section 8.3.1.

Since the environment is largely outdoors, GPS was initially considered for localisation. Unfortunately, the robot is required to operate in and around metal structures and under trees, rendering GPS insufficiently reliable.

An initial approach to using the laser for localisation was to build a prior occupancy grid [37], then to localise based on measured ranges to obstacles. This technique has been shown to be extremely robust in indoor environments [110]. Particles are weighted using a likelihood function based on the differences between the ranges of actual returns and the expected ranges of returns, where the expected ranges are calculated by ray-tracing in the occupancy grid.

The occupancy grid approach was non-trivial to apply to the problem at hand because the laser often sees the ground. This is rare in indoor environments because the ground is generally flat and the laser scan is generally close to horizontal. In the present scenario, the terrain contains small bumps and gradients, and the dynamically-stabilised platform needs to pitch back and forth to remain upright. When accelerating or even just leaning into the wind, the robot often sees the ground only a few metres in front of it. An example is shown in Figure 8.3.

While it may be possible to model the interaction with the ground, the solution adopted was to extract point features from the laser scan. The most reliable way of doing this was to modify the environment, adding laser-reflective fiducials as shown in Figure 8.1. The likelihood function

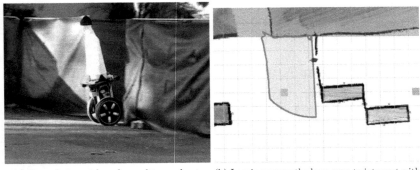

(a) The robot must lean forward to accelerate   (b) Leaning causes the laser scan to intersect with the ground

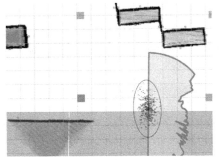

(c) This intersection is difficult to model because the ground is not perfectly flat

Figure 8.3: An example showing the impracticality of modelling the world using a two-dimensional occupancy grid. The robot often needs to lean forwards at a considerable angle in order to accelerate, as shown in (a). (b) shows the associated laser scan in red, drawn from the mean pose. It intersects the ground approximately 6m in front of the robot. If the ground were perfectly flat, it would be possible to use information from the robot's gyroscopes to calculate the expected point of intersection. Unfortunately it is generally not flat. (c) shows one example of un-even terrain in front of the robot.

can then be constructed based on the ranges and bearings to the actual and expected sets of observed features. The expected set of observed features can be assessed by ray-tracing through the prior occupancy-grid map. Assuming that the fiducials are the only retro-reflective objects in the environment, the robot may fail to detect true features, due to factors such as the tilting of the platform, but it is unlikely to register a false positive. It is clearly not a requirement of the PPOMDP algorithm in general that an environment be instrumented, however it simplifies matters considerably in this domain.

8.3 From BlockWorld to the Real World

8.3.1 The PPOMDP Model

The previous section provided an outline and motivation for the method used for estimation. This section gives precise details of the model used for simulating the world during planning and for estimation.

States and Beliefs

The state of the world consists of an occupancy grid, a set of features, and the state of the robot. The occupancy grid and features are considered to be known and static. The set of features is of the form $F = \{\mathbf{f}_i | i = 1 \dots |F|\}$, where $\mathbf{f}_i$ is the tuple $< x_i, y_i >$ defining the fixed position of the i'th feature.

The state-space of the robot is defined by the tuple $\mathbf{x} = < x, y, \theta >$ describing the robot's pose. The parametric belief-space is the space of Gaussians defined by 9 parameters: 3 describing the mean and 6 describing the upper triangle of a full 3x3 covariance matrix (due to symmetry, only the upper triangle needs to be represented explicitly).

Actions

The action-space is defined by the tuple $\mathbf{u} = < v_l, v_\theta >$, where v_l and v_θ are the linear and rotational velocities respectively. The range of valid input values for v_l is $[0, 2]$ metres per second, and for v_θ is $[-90, 90]$ degrees per second. During planning, time is discretised to intervals of $\Delta t = 1$ second. Therefore for a single action the maximum nominal linear change in position is 2 metres, and the maximum nominal change in heading is 90 degrees.

The Process Model

For a given nominal action $< v_l, v_\theta >$, the true (noisy) linear and angular velocities, v_l' and v_θ', are distributed according to independent Gaussians $\mathcal{N}(v_l, (0.25v_l)^2)$ and $\mathcal{N}(v_\theta, (0.35v_\theta)^2)$ respectively, where $\mathcal{N}(\mu, \sigma^2)$ denotes a Gaussian with mean μ and variance σ^2. This noise model was determined through extensive empirical testing with the robots used to demonstrate this work. In the absence of obstacles, the planning model applies an action by first translating the robot by $v_l'\Delta t$ in the direction of its current heading then rotating it by $v_\theta'\Delta t$. Where an action would traverse a non-empty cell of the occupancy grid, the model applies no change to the pose.

Observations

An observation consists of the tuple $\mathbf{z} = <\mathbf{z}_C, Z_f>$. $\mathbf{z}_C$ is an observation of whether or not the previous action was successful, identical to the collision sensor used in BlockWorld. $Z_f = \{\mathbf{z}_{f,i}|i = 1\ldots|Z_f|\}$ is a set of observations of features, where $\mathbf{z}_{f,i}$ is the i'th range-bearing tuple $< z_{r,i}, z_{b,i} >$. Since the laser is mounted forward and has a $180°$ field of view, $z_{b,i}$ is in the range $[-90°, 90°]$, where a bearing of zero is directly in front of the robot.

While the maximum range of the laser is 80m, features are not observable at this range. The tilting of the platform and slope of the ground mean that the laser scan often does not intersect the fiducials at longer ranges. Even when the scan does intersect the fiducials, the laser has difficulty detecting the high reflectivity beyond relatively short ranges. Due to the combination of these effects, the maximum range at which features can reliably be detected is approximately 8m. Building a probabilistic model beyond 8m is problematic because the observability of features is state-dependent, given small variations in the slope of the ground in different areas of the environment. Therefore, as a slight simplification, it is assumed that the maximum range of the laser is exactly 8m, and hence $z_{r,i}$ is in the range $[0, 8]$. To ensure that this simplification matched reality, the laser was artificially limited to an 8m range during live execution.

The Sensor Model

Let the expected set of feature observations from a particular pose be denoted $\hat{Z}_f$. $\hat{Z}_f$ is determined by first calculating, for each $\mathbf{f}_i \in F$, the expected range and bearing to that feature, denoted $< \hat{z}_{r,i}, \hat{z}_{b,i} >$. If $\hat{z}_{r,i}$ and $\hat{z}_{b,i}$ are within the bounds defined above for valid observations, and a ray from the vehicle to the feature does not intersect an occupied cell of the occupancy grid, $< \hat{z}_{r,i}, \hat{z}_{b,i} >$ is added to $\hat{Z}_f$.

The sensor model assumes that true observations are a perturbation of the expected observations. It assumes that the range and bearing of the j'th actual feature observation, $< z_{r,j}, z_{b,j} >$, are drawn from the Gaussian distributions $\mathcal{N}(\hat{z}_{r,j}, (0.3\text{m})^2)$ and $\mathcal{N}(\hat{z}_{b,j}, (5°)^2)$. In addition to sensor noise from the laser, these noise levels account for inaccuracies in timing (particularly during sharp turns), inaccuracies in the map, inaccuracies in the measured transformation from the vehicle reference frame to the laser, and the fact that laser rays are not always exactly horizontal.

The Likelihood Function

The likelihood function used for re-weighting particles is a function of two observations: the true observation, $\mathbf{z}$, and the expected observation given the particle, $\hat{\mathbf{z}}$. It is the product of two factors

$$l(\hat{\mathbf{z}}|\mathbf{z}) = l_C(\hat{z}_C|\mathbf{z}_C)l_f(\hat{Z}_f|Z_f) \tag{8.1}$$

where the factors l_C and l_f are functions of the collision and feature components of the observation respectively. For the binary collision component, $l_C(\hat{z}_C|\mathbf{z}_C)$ returns zero for a mismatch, and one otherwise.

The feature-based component l_f is more complicated, since one must consider not only sensor noise but also the likelihood of non-detection of features. Furthermore, there are two scenarios in which the likelihood function may be applied: the simulation used for planning, and the real world. In the former, observations occur at a frequency of 1Hz, whereas in the real world observations occur at the frequency of the laser, which runs at about 10Hz.

To deal with these issues, the assumption is made that during a given one-second interval, the probability of failing to observe an observable feature is negligible. The likelihood function used for planning therefore assumes that the probability of non-detection is zero, and operates as follows. Of the actual and observed sets of features, let Z_{f+} and Z_{f-} denote the larger and smaller sets respectively ($|Z_{f+}| \geq |Z_{f-}|$). If $|Z_{f-}| = 0$ but $|Z_{f+}| \neq 0$, l_f returns zero. Otherwise, l_f is calculated using the product

$$l_f(\hat{Z}_f|Z_f) = 1.0 \times \prod_{i=1}^{|Z_{f+}|} l_s(\mathbf{z}_{f+,i}|Z_{f-}) \tag{8.2}$$

where l_s is a likelihood function for single feature observations. $l_s(\mathbf{z}_{f+,i}|Z_{f-})$ is the likelihood of the i'th feature observation from Z_{f+} given the set of feature observations Z_{f-}. It is based on the similarity of $\mathbf{z}_{f+,i}$ to the closest-matching feature observation in Z_{f-}:

$$l_s(\mathbf{z}_f|Z_f) = \max_{\mathbf{z}_{f,j} \in Z_f} G(z_r - z_{r,j}, \sigma_r^2)G(z_b - z_{b,j}, \sigma_b^2) \tag{8.3}$$

where G is the Gaussian function

$$G(x, \sigma^2) = \frac{1}{\sqrt{2\pi\sigma^2}} exp(\frac{-x^2}{2\sigma^2}) \tag{8.4}$$

This likelihood function is not appropriate during online estimation because it ignores the possibility of non-detection. For online estimation, a slightly different likelihood function $\tilde{l}_f$ is used, given by

$$\tilde{l}_f(\hat{Z}_f|Z_f) = \nu(\hat{Z}_f|Z_f) \prod_{i=1}^{|Z_f|} l_s(\mathbf{z}_{f,i}|\hat{Z}_f) \tag{8.5}$$

where $\nu(Z_f|\hat{Z}_f)$ is a penalty term which penalises expected observations against which no actual observations were matched. If the number of non-matched observations is n, ν is chosen to be

0.75^n. In other words, for every observation which a particle predicts but which is not actually observed, the likelihood is multiplied by 0.75. The figure 0.75 was experimentally found to produce consistent beliefs.

The Reward Function

The environment contains a goal area and one or more hazard zones. The reward as a function of state is +10 for being in a goal area, −50 for being in a hazard zone, and −0.1 otherwise. As for the BlockWorld problem, the reward as a function of belief is calculated by integrating the agent's belief over the areas of interest. Integrations are performed using sampling.

8.3.2   Discussion of the Model

Before continuing, it is worth assessing the difficulty of the RealWorld problem relative to the BlockWorld problem. The length of the state vector has been extended from two to three with the addition of heading, and full 3×3 covariance matrices are being modelled rather than diagonal 2×2 matrices. The length of parameter vectors is therefore increased from 4 to 9. This represents a huge leap forward in complexity. For example, using a regular grid of beliefs for B is clearly intractable. Assuming that the heading is discretised into 10 bins and that each element of the covariance matrix is discretised into 6 levels (as was the case in Chapter 3), a regular grid for the current problem would have approximately 13,000 times more belief-points than a regular grid for a BlockWorld problem of the same physical size.

Using an arbitrary set B improves matters, however the achievable density of belief-points will certainly be lower than was possible for the BlockWorld problem. Chapters 6 and 7 experimented with various settings for the size of B. The results showed that performance increases rapidly as the density of belief-points increases, then reaches a plateau at some density. The extra dimensionality of the belief-space for this more realistic problem suggests that it may not be possible to operate on this plateau. As discussed in Chapter 7, the use of forward planning was particularly helpful when operating at a portion of the curve prior to the plateau. It is anticipated that forward planning will be similarly useful for this problem.

Relative to previous work on POMDP-based robot navigation, we believe that the problem described above is substantially more realistic than any robot navigation problem which has been attempted to date. To the author's knowledge, the most similar problems from the literature are (a) the navigation problem presented by Spaan and Vlassis, using omni-directional vision [103], and (b) the navigation problem presented by Roy [92], using back-to-back SICK lasers in a realistic simulation. Both of these problems assumed heading-invariant sensors and modelled only robots' (x, y) positions during planning. In the real world, robots do have headings and most robots also have directional sensors. A planner with no concept of heading can have no understanding of the fact that heading uncertainty induces positional uncertainty, and cannot account for non-holonomic constraints [63]. Allowing only omni-directional sensors prohibits robots from exhibiting the interesting and useful behaviour of moving in order to point their sensors in informative directions.

8.3.3 PPOMDP Parameter Settings

This section lists the settings which were used for the PPOMDP algorithm when applied to more realistic problems. Essentially, they represent a generalisation of the settings which were found to be most successful on the BlockWorld problem.

Observation Pre-Calculation

The action-independent portion of the observations were pre-calculated in a similar way as they were in the BlockWorld problem, as described in Chapter 5. Observations were split into two components, z_C and Z_f, of which only the former is action-dependent. Belief transitions can therefore be split into two sub-transitions: T_C (resulting from the action and collision observation) and T_f (resulting from the feature observations).

Belief Sets and Weighting Functions

Arbitrary belief sets were used. Chapter 6 defined a number of new parameters which must be specified in order to use an arbitrary belief set. The following choices were made:

1. **Similarity Search**: vp-tree as the similarity-search algorithm;

2. **Distance Metric**: the Euclidean distance metric in parameter space D_{EP} was used, measuring distances in metres and angular quantities in radians;

3. **Kernel Bandwidths**: half the mean distance to the closest point in the belief set, as described in Chapter 6;

4. **Belief Sets**: a single set was used for both the beliefs after full observations (B) and the beliefs after partial observations (B_C). The set was chosen using exploration as described below.

Exploration for belief set generation was performed by repeatedly (a) randomly initialising the state at a non-occupied position in the map, (b) initialising a belief, consistent with that state, with a small diagonal covariance, then (c) executing 30 actions, adding each posterior belief to B. This process was continued until the belief set was of a specified size.

The policy used to explore was a mixture of random and heuristic policies. At each step, with probability 0.75 a random action was taken, selected from a uniform distribution over the space of valid linear and angular velocities. With probability 0.25 the action was taken using the non-probabilistic policy which will be described in Section 8.4.3.

As was discussed in Chapter 6, the optimal belief set should have a belief density corresponding to the probability of occurrence of beliefs when executing the final policy. Since the final policy cannot be known before the belief set is chosen, an approximation to that policy is needed. A mixture of random and non-probabilistic policies was chosen as this approximation: the PPOMDP and non-probabilistic policies are likely to coincide often, however using a significant random component allows the PPOMDP planner to consider plans which the non-probabilistic policy would not make.

8.4 Evaluation of Policies

Policies were evaluated in three scenarios of increasing complexity and realism:

1. the simple simulator used for planning;

2. a realistic simulator; and

3. the real world.

Previous chapters have considered only the first scenario. This section describes the framework used for evaluating policies in the latter two scenarios, the process noise model used for realistic simulation, and the non-probabilistic policy against which PPOMDP is compared in all three scenarios.

8.4.1 The Component Framework for Online Evaluation of Policies

The framework used for evaluating policies online, both in realistic simulation and the real world, is shown in Figure 8.4. The system was implemented such that the exact same software could control either a simulated robot or a real robot, using Orca [20]: an open-source software framework for building component-based robotic systems. Components for hardware/simulation interfacing, feature extraction, obstacle avoidance, and visualisation are available for download from the Orca component repository[1].

---

[1]http://orca-robotics.sourceforge.net

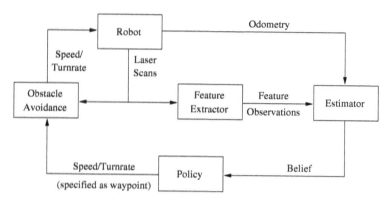

Figure 8.4: The component framework used for evaluating policies during realistic simulation or live execution.

The robot receives speed and turn-rate commands from a low-level obstacle avoidance algorithm, and produces laser scans and odometry information. A feature extractor processes the laser scans to segment the fiducials. An estimator (or localiser) receives both the extracted features and the odometry information, producing Gaussian beliefs. The policy then maps from beliefs to actions.

A low-level obstacle avoidance algorithm mediates between the policy and the robot. This is done because the policy maps directly from a belief to an action, without reference to local sensing. Since the estimator relies on extracted features rather than occupancy, the policy may have no way of knowing that there is an obstacle directly in front of it. The obstacle avoidance algorithm used in this case was VFH+ [117].

One complication is that the VFH+ accepts waypoints rather than direct commands of speeds and turnrates. It converts these waypoints into speed and turnrate commands depending on local sensing. This complication was handled by having the policy set constraints on VFH+'s maximum speed and turnrate in addition to setting waypoints. Fine control is possible by setting the maximum speed and turnrate to the desired values while issuing waypoints only a small distance in front of the robot. VFH+ will then execute the desired speed and turnrate when there are no obstacles in the vicinity. While all components run asynchronously at frequencies appropriate for their tasks, the policy sends commands to the obstacle avoidance component at 1Hz (the Δt used during planning).

8.4.2 The Process Noise Model Used in Realistic Simulation

Since policies are to be evaluated in realistic simulation first, a model is needed to inject odometric noise into that simulation. It is common in mobile robot localisation problems to assume additive white Gaussian odometric noise. We argue that in real scenarios this assumption is almost always invalid, but it is made because it is sufficiently close to the truth. In reality, for any given model, systematic errors occur due to issues such as non-uniform terrain and un-modelled dynamics. To account for these systematic biases, the usual approach is to increase the level of assumed Gaussian noise.

In an effort to be as realistic as possible, we attempt to re-create this scenario. The model for localisation assumes that actions are perturbed by independent identically-distributed Gaussian noise sampled from $\mathcal{N}(1, (0.25v_l)^2)$ and $\mathcal{N}(1, (0.35v_\theta)^2)$ for linear and rotational components respectively, as described in Section 8.3.1. The true noise injected into the simulation, however, is neither independent nor Gaussian. Rather, a pair of multipliers m_l and m_r are sampled from uniform distributions over the ranges $[-0.25, 0.25]$ and $[-0.35, 0.35]$ respectively. The true additive noises are then deterministically set, at each time-step, to $m_l v_l$ and $m_r v_\theta$. Every 10 seconds, a new pair of multipliers is sampled.

Relative to the model, the true simulated noise is relatively small in magnitude. The true odometry is always less than or equal to one standard deviation from the nominal value. Indeed, the localiser has no difficulty in coping with the noise, despite the fact that the model is strictly incorrect. The advantage of using non-independent noise is that it gives realistic errors, such as odometry sometimes being incorrect by a significant amount for large sharp turns. Section 8.6 will present results which indicate that the level of noise used in simulation is not unrealistic. If anything, it is relatively tame compared to the noise experienced when operating in a real environment.

8.4.3   The Competition: Non-Probabilistic Path Planning

The PPOMDP-based controller was evaluated by comparing against the performance of a simple non-probabilistic policy, which we refer to from this point onwards as NONPROB. At a frequency of 1Hz NONPROB computes a deterministic path, as a series of waypoints, from the maximum-likelihood position to the goal. This is identical to the *Replan* heuristic discussed in Section 2.7.1, except that NONPROB re-plans on every iteration rather than just when it detects that its plan has gone awry. This is not problematic, since deterministic re-planning is computationally inexpensive.

The description of NONPROB requires the specification of how paths are planned and how they are followed. Two different algorithms are used for path-following: one during execution in the simulator used for planning, and another during realistic simulation or live execution. Both path generation and path following are described below.

Path Generation

Paths are calculated using well-known techniques from the motion-planning literature [63][64]. The occupancy grid map is first pre-processed by

1. marking all cells within hazards as occupied;

2. growing occupied cells by the radius of the robot;

3. calculating a potential function, over all of free-space, directly proportional to the distance to the nearest obstacle; and

4. extracting a "skeleton" corresponding to the loci of the maxima of this potential function.

The skeleton is a path which connects areas of free-space while keeping a distance from obstacles. From a start point, planning a path to a goal point involves

1. connecting to the closest point on the skeleton,

2. traversing the skeleton to the skeleton-point nearest to the goal, and

3. connecting to the goal point.

As a post-processing step the path is optimised by removing many waypoints while still maintaining clear straight lines between all waypoints. For more detailed information, the reader is directed to [63]. The code to perform this planning is also available from the Orca component repository.

Path Following in the Planning Simulator

When using the planning simulator, NONPROB translates beliefs and paths into actions using simple rules. If the difference between the maximum-likelihood heading and the heading to the next waypoint is more than $20°$, the policy turns towards the next waypoint. It sets a linear velocity of $0.5m/s$ and turns towards the waypoint as fast as possible, without setting the rotational velocity so high as to overshoot the desired heading in the one second alloted for the action. If the difference in heading is less than $20°$, the policy approaches the next waypoint. It sets the linear velocity as high as possible, avoiding overshoot, while setting a rotational velocity that will keep the maximum-likelihood heading pointed towards the waypoint.

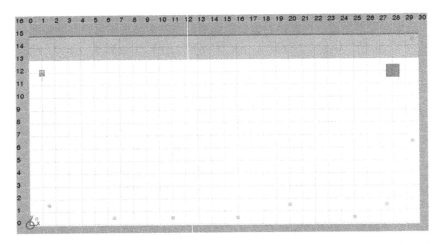

Figure 8.5: The toy world. The dark yellow area at the top is a hazard which the robot should not enter. The robot icon near the top left marks the start position, which faces south. The 1m×1m goal is marked by the dark cyan square near the top right. The small red squares are features. The map measures 30m×15m. Occupancy grid cells measure 0.1m×0.1m. in this case, all cells are empty.

Path Following Online and During Realistic Simulation

The path-following algorithm used during online plan execution or realistic simulation is also simple. As shown in Figure 8.4, VFH+ obstacle avoidance mediates between the policy and the robot. NONPROB simply gives the set of waypoints generated by the deterministic plan to VFH+, relying on VFH+ to follow the path under the assumption that the maximum-likelihood state is the true state. VFH+ is commanded to reach the final waypoint (on the goal) with zero tolerance, such that it persists until the episode ends, whether through success, failure, or expiration of the allotted time.

8.5   ToyWorld: A Toy Problem with Realistic Dynamics

Before embarking on RealWorld, the model described above is applied to ToyWorld: a toy problem, but with realistic dynamics. The simulation was implemented using the Stage simulator [48]. Figure 8.5 illustrates ToyWorld.

For this world, the obvious strategy from the start location is to turn and drive directly towards the goal. Indeed, this is exactly the strategy taken by NONPROB. It pushes the mean of its distribution towards the goal, ignorant of the amount of probability mass that may be sweeping

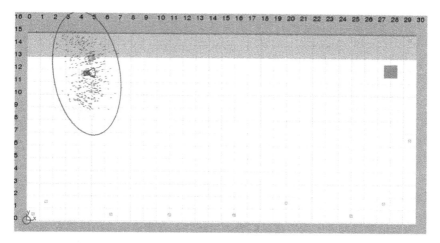

Figure 8.6: An example distribution encountered during non-probabilistic control. The particles are shown in blue, and the parametric belief is indicated by the blue ellipse. The true state, shown in cyan, is north of the mean due to oversteer on the initial left turn from the start point. Since the non-probabilistic controller is aware of only the mean of the distribution, it is completely unaware that there is a danger. After the next action, the robot will have entered the hazardous area.

through the hazardous area. An example distribution that arises using non-probabilistic control is shown in Figure 8.6.

In contrast, a POMDP-based controller is able to consider the entirety of its distribution. It should realise that it is incapable of turning by exactly 90° from the start point, and that there are no features near the top of the map which can be used to correct any errors. A better strategy is therefore to begin by moving south towards the features, traverse eastwards using the features to navigate, then finally move north to the goal. The path is longer but more reliable.

8.5.1 Results

The PPOMDP policy was generated using a belief set containing 4000 beliefs. 200 samples were used to estimate the belief transition function during planning. This number was reduced during online forward planning: 100 samples were used at the root of the game-tree. The average time required to make a decision while executing the plan was approximately 95ms. The total time taken to generate the value function was approximately 420 seconds, or 7 minutes. The breakdown of this figure is shown in Table 8.1. This breakdown indicates that belief set generation is by far the most time-consuming component of value function generation,

Step	Required Time
Belief Set Generation	325.5s
Observation Precalculation	59.7s
Calculation of MDP Transitions	32.1s
Reward Calculation	0.4s
Discrete MDP Solving	2.9s
Total	**420.6s (7min)**

Table 8.1: A breakdown of the time required for PPOMDP to generate a value function for the toy world.

followed by observation precalculation and MDP transition calculation. A large component of all three is ray-tracing. Planning could therefore be accelerated by the use of a data structure which allows fast ray-tracing, such as a quad-tree [96]. Another possibility is to pre-compute the ray-tracing, as suggested by other authors [43]. These optimisations were not exploited in this book.

Results Using the Planning Simulator

Policies were evaluated in the planning simulator over 100 episodes. An episode is terminated when the robot reaches the goal or enters a hazard, or 100 seconds have elapsed. The mean reward for PPOMDP was 3.8 compared with -22.3 for NONPROB. NONPROB's reward corresponds to an approximate success rate of 50% (the mean of +10 for success and -50 for failure, minus a small amount for the time spent reaching the termination condition). This is not unexpected: NONPROB is as likely to oversteer as understeer on the initial turn. Oversteer will lead to failure, understeer to success.

Results Using the Realistic Simulator

Policies were evaluated over 40 episodes on the realistic simulator, using the same world described previously. The simulation was reset at the end of each episode. PPOMDP achieved a perfect success rate, for an average reward of 6.2, while NONPROB failed on 15 of the 40 episodes, for an average reward of -14.8. In order to contrast the strategies adopted by the two policies, Figure 8.7 plots the true trajectories taken over the 40 trials, for each controller.

From Figure 8.7(a), it can be seen that when NONPROB oversteers on the initial turn, it enters the hazard. If it understeers it will eventually observe a feature, adjust its heading, and reach the goal. PPOMDP adopts an entirely different strategy, using the features to allow it to reliably reach the goal. One interesting point is the behaviour near the goal, where the robot sometimes circles once before the end of the episode. This is likely due to a modelling

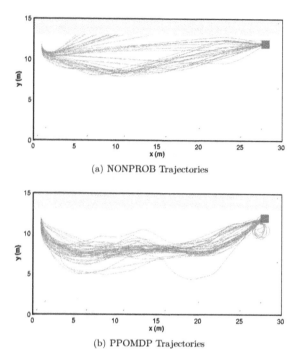

(a) NONPROB Trajectories

(b) PPOMDP Trajectories

Figure 8.7: The true trajectories taken over 40 trials during realistic simulation in ToyWorld, using (a) the non-probabilistic policy and (b) PPOMDP. The start point is on the left, while the goal is on the right. Goals, hazards and features are marked in blue, yellow and red respectively.

approximation. The process model used for planning, described in Section 8.3.1, applies the entire translation before applying the entire rotation associated with an action. Therefore the robot believes that by turning right while moving forward, it will arrive directly in front of its current position, only with a different heading. Instead it arrives somewhat to the right of its current position, sometimes missing the goal on the first attempt.

Another point of interest is that PPOMDP's trajectories are significantly more smooth than NONPROB's. This is probably due to the interaction between PPOMDP and the VFH+ obstacle avoidance component. As described in Section 8.4, PPOMDP tries to set speeds and turnrates using an *ad hoc* scheme involving setting waypoints and constraints. It is likely that this scheme does not allow PPOMDP to turn as tightly as it would like. In contrast, NONPROB selects much more sparse waypoints, leaving VFH+ with the control to make sharp turns.

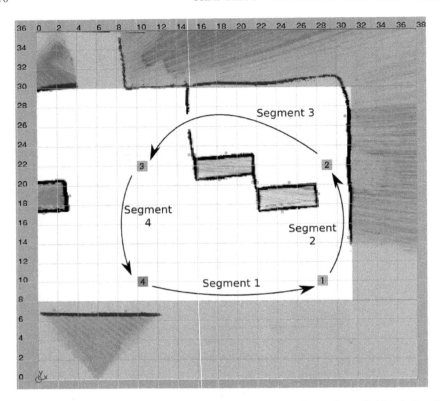

Figure 8.8: The map used for the RealWorld problem. The colour scheme is identical to the scheme used in Figure 8.5. The numbers on the goals indicate the order in which they should be traversed (counter-clockwise). The four segments are indicated by the arcs. The map measures 38m×36m, with 0.1m×0.1m occupancy grid cells.

8.6 The RealWorld Problem

8.6.1 Problem Description

The RealWorld problem uses the right half of the map from Figure 8.2(d). Since there will be no simulation to reset when running live, it is difficult to transport the robot back to the starting location after each episode. The robot is therefore given four goals, as shown in Figure 8.8. Its task is to reach the four goals in order, in a counter-clockwise direction, starting from the lower-left goal. This brings it back to the starting location for the next episode.

Multiple Goals

The use of multiple goals introduces a problem, since the value function depends on the reward function, which is specific to a single goal. As a solution, four value functions are generated: one per goal. When the robot reaches a goal it simply loads the value function for the next goal.

Of the steps required to produce a value function, several can be re-used for multiple goals. Specifically, only one belief set needs to be generated, and observation precalculation and MDP transition calculation need to be performed only once, since they are independent of reward. Since these steps dominate the total time required, as shown in Table 8.1, the cost of adding extra goals is small.

One complication with re-using belief set generation is that the heuristic policy used to bias random exploration, as described in Section 8.3.3, requires a goal. This problem is solved by using each goal for one quarter of the belief set generation process.

Tunnelling Through Thin Walls

A second complication, which had not been encountered previously, arises from the fact that the walls in RealWorld are thin. The approach to planning described thus far can lead PPOMDP to believe that it can 'tunnel' through thin walls in two ways.

Firstly, the planner maps all distributions to the nearest Gaussian, and therefore cannot consider distributions involving a step change. Step changes do occur in RealWorld, however, where uncertain distributions meet walls. Since Gaussians have smooth tails, the planner believes that if it becomes sufficiently uncertain and moves the mean of its distribution near a thin wall, there will be a non-zero probability that the true state is in fact on the opposite side of that wall. It believes that the right feature observation (*i.e.* one that is only possible after tunnelling through the wall) will re-weight its distribution such that it becomes well-localised on the opposite side of the wall. Of course in reality such an observation is impossible since the probability that the robot has passed through the wall is zero. However, during experimental trials, PPOMDP clearly tried to take advantage of this loophole. It intentionally avoided looking at features while hugging thin walls, hoping to tunnel through.

Secondly, even in the absence of observations from the opposite sides of walls, the planner may try to take advantage of the "snapping" of posterior beliefs onto B. The distance measure considers only the proximity of beliefs, not the the occupancy grid. Especially in the presence of a locally sparse belief set, a nearby belief in B may be on the opposite side of a thin wall, leading the planner to believe that it can tunnel through.

One solution is to use the occupancy grid to modify the weighting function. If direct line-of-site does not exist between the mean of a posterior and the mean of a member of B, that member

Step	Required Time
Belief Set Generation	568.5s
Observation Precalculation	697.4s
Calculation of MDP Transitions	178.2s
Reward Calculation (per goal)	0.7s
Discrete MDP Solving (per goal)	9.2s
Total	**1483.7s (~25min)**

Table 8.2: A breakdown of the time required for PPOMDP to generate a value function for the RealWorld problem. The last two items are multiplied by four in the calculation of the total, since they must be performed once per goal.

of B must receive zero weight. This solution is rather *ad hoc*, but worked for this problem. There may be circumstances where it rejects legitimate nearby beliefs (such as near corners), but the effects were not noticeable. All RealWorld results were generated with the inclusion of this feature.

8.6.2 Results

Computation Time

The size of the belief set was chosen to be 8000 beliefs. The number of samples used to estimate the belief transition function was increased to 500, although this is almost certainly excessive. As for ToyWorld, this number was decreased to 100 during forward planning.

The average time required to make a decision during plan execution was approximately 145ms. This is longer than the time required for ToyWorld for three reasons. Firstly, the larger belief set induces a sub-linear increase in the time required for each similarity search when applying the weighting function. Secondly, the increase in the number of features causes a corresponding increase in the time required to calculate expected observations. Thirdly, since the world is larger, ray-traces for observations are likely to cover more distance on average.

The total time required for planning all four segments of the circuit was 1484 seconds, or approximately 25 minutes. The breakdown of this total is shown in Table 8.2. Compared to ToyWorld, observation precalculation and MDP transition calculation consume larger portions of the total time. This is because the increase in the number of samples (from 200 to 500) affects only these two phases. The number of samples used during belief set generation is controlled by the KLD particle filter, which adapts the number of samples.

Segment	NONPROB	PPOMDP
1	-18.7	1.4
2	8.5	5.9
3	4.9	8.1
4	9.0	8.9

Table 8.3: The average reward obtained by either policy on each of the four segments of the loop around RealWorld.

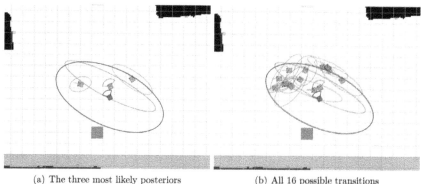

(a) The three most likely posteriors   (b) All 16 possible transitions

Figure 8.9: The perceived distribution over future posteriors in B, over a one-step planning horizon. The current belief is marked in blue. The considered action is to move forward while turning slowly to the right. Each green ellipse is a belief in B which the planner considers to be a possible result of this action. (a) shows the three most likely posterior beliefs. From left to right, the transition probabilities are 0.05, 0.8, and 0.03. The posteriors correspond to observing the feature on the left, observing nothing, and observing the feature on the right, respectively. Note that the posterior corresponding to no observation has a large heading uncertainty. (b) shows all 16 possible posteriors. This shows how each slightly different possible feature observation will produce a slightly different belief, whereas the absence of an observation will result in one poorly-localised belief.

Results Using the Planning Simulator

The policies were initially compared using the planning simulator. Each policy attempted each of the four segments of the loop 100 times. The results are shown in Table 8.3. The most interesting segment is the first, where PPOMDP clearly outperforms NONPROB. The reason for this difference will be explained with reference to the results obtained using the realistic simulator. To help develop intuition for the operation of the algorithm, Figure 8.9 shows a snapshot of the forward planner's perceived distribution over future posteriors.

Segment	Average Reward		Number of Failures		Number of Successes	
	NP	PP	NP	PP	NP	PP
1	-6.7	4.3	10	1	30	39
2	8.6	7.9	0	0	40	40
3	7.6	6.9	0	0	40	40
4	8.7	8.1	0	0	40	40

Table 8.4: A comparison of the performance of the two policies on each segment of the Real-World problem during realistic simulation. NONPROB and PPOMDP are abbreviated to NP and PP respectively.

Results Using the Realistic Simulator

Using the realistic simulator (Stage [48]), policies were evaluated over 40 loops. During execution, if a segment was terminated by either a hazard or the expiration of time, the segment was considered a failure and the simulation was reset at the next goal, facing away from the previous goal. The results are shown in Table 8.4, and the true trajectories are plotted in Figure 8.10.

NONPROB's failure mode is clear from Figure 8.10(a). Understeer while turning at the lower-left goal causes the robot to enter the hazard to the south. To avoid this, rather than turning 90° left after reaching the lower-left goal, PPOMDP usually wheels 270° to the right in a large arc. This allows the robot to see the feature just below the lower-left goal during the first part of the turn, then finish the turn by seeing either the features to the centre-left or near the centre of the map. The result is that the robot reliably reaches the area near the lower-right goal.

Again, process model approximations cause the robot to sometimes circle just prior to reaching the goals, particularly near the lower-right and upper-left goals. PPOMDP's one failed trajectory can be seen, just below the lower-right goal. One possible reason for this is particularly poor odometry while circling to reach the lower-right goal. Another possibility is temporary localisation failure, which did occur occasionally during execution.

On the other three segments of the loop, it can be seen from Table 8.4 that NONPROB slightly out-performs PPOMDP. While the circles before reaching goals account for some of this difference, NONPROB also clearly takes a more direct approach than PPOMDP, conferring a small advantage in the absence of uncertainty. Again, this is likely a result of the interaction between PPOMDP and the obstacle avoidance component.

Results From Running Live

Finally, both policies were evaluated on real robots. Since the true pose of the robot is unavailable, different termination criteria are needed. The robot is deemed to have reached a goal if it places the mean of its belief over that goal while its uncertainty is sufficiently low (the variance in x and y must be less than one metre). The uncertainty condition is trivial to satisfy if the

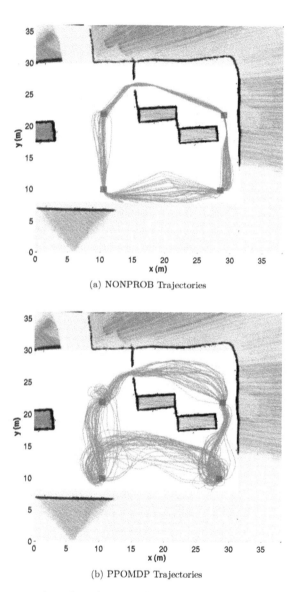

(a) NONPROB Trajectories

(b) PPOMDP Trajectories

Figure 8.10: The true trajectories taken over the 40 trials during realistic simulation in Real-World, using (a) the non-probabilistic policy and (b) PPOMDP. Goals, hazards and features are marked in blue, yellow and red respectively.

	NONPROB	PPOMDP
Number of Successful Loops	3	27
Number of Failed Loops	8	4
Total attempted Loops	11	31
Success Rate	**27%**	**87%**

Table 8.5: The number of successful and failed loops using the two policies on a real robot.

robot is not lost, since the goals are all in sight of features. Termination due to hazards cannot be assessed in the same way, because the robot will never knowingly enter a hazard. Instead this was assessed manually, by watching the robot.

The results are presented in Table 8.5. Figures 8.11 and 8.12 show snapshots of typical behaviour, under NONPROB and PPOMDP control respectively. Numerically, PPOMDP's performance is clearly superior to NONPROB. Subjectively, it appears that NONPROB's problem is that it is over-confident. This is much more apparent during live execution on a large robot than in simulation. NONPROB, with its complete faith that its maximum-likelihood position is correct, seems like a dangerous driver and is stressful to watch. In contrast, PPOMDP appears far more cautious and intelligent, taking care to ensure that it is well-localised before acting.

Table 8.5 shows only loops that were attempted in their entirety. A handful of loops were aborted due to policy-unrelated issues, such as failure of the localisation module, or collision with an obstacle. The latter occurred several times when trying to pass through the door near the top of the map. An unexpected advantage of PPOMDP was that it was far less likely to have problems with this door. A possible reason for this is that the model used for planning does not include obstacle avoidance, but rather assumes that an attempt to move through the door when incorrectly aligned will simply fail. This impacts visibly on PPOMDP's behaviour: it slows down when approaching the door, receiving more observations per metre travelled than NONPROB does, and hence is better localised as it passes through the door.

An interesting issue is that the results in Table 8.5 are much worse than the results obtained in simulation. The fact that NONPROB failed on significantly more than 50% of loops suggests that the odometry is biased to understeer on the left turn near the lower-left goal. One possible explanation is that the odometry is always biased to the right. A more likely explanation is that, as the robot corners hard at speed after reaching the lower-left goal, the weight is thrown onto the outside tyre which compresses. The outside wheel therefore has a reduced effective radius, and travels less distance for the same angle of rotation than the inside wheel. These kinds of dynamics are difficult to model, justifying the approach of simply using more (assumed Gaussian) process noise.

Another major difference between simulation and reality is in the observation model. The planning simulator guarantees feature observations at 1Hz if there is line-of-sight to a feature

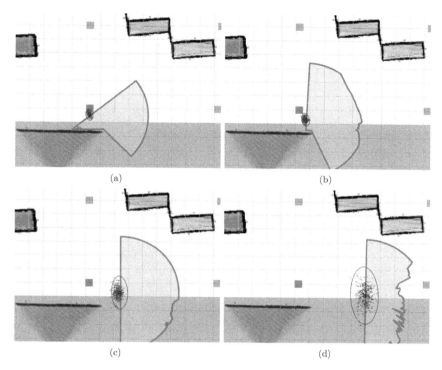

(a)   (b)

(c)   (d)

Figure 8.11: Screenshots during live execution, showing NONPROB's typical behaviour. After reaching the lower-right goal, NONPROB turns to the left as shown in (a). It soon loses sight of all features, as shown in (b). As a result, uncertainty quickly grows as it travels towards its goal. NONPROB considers the belief in (c) to be perfectly safe because the maximum-likelihood pose is not in the hazard to the south. This is despite the fact that considerable probability mass exists within that hazard. By the time the robot reaches the belief shown in (d), the true pose has entered the hazard and the robot sees the uneven ground in front of it.

less than 8m away. The realistic simulator also guarantees feature observations, but at an unrealistically high frequency. It was anticipated that, of the two, the planning simulator would be the more realistic model for reality. However, in reality observations are not guaranteed at 1Hz. Often, a second elapses without an observation of a feature within range, for example due to the tilting of the platform. A more accurate model would be much more complicated. The transition from the planning simulator to the realistic simulator maintains a simple observation model but represents a significantly more complicated process model. The transition from the realistic simulator to reality adds a significantly more complicated observation model.

Most of PPOMDP's four failures involved wandering south into the hazard when very uncer-

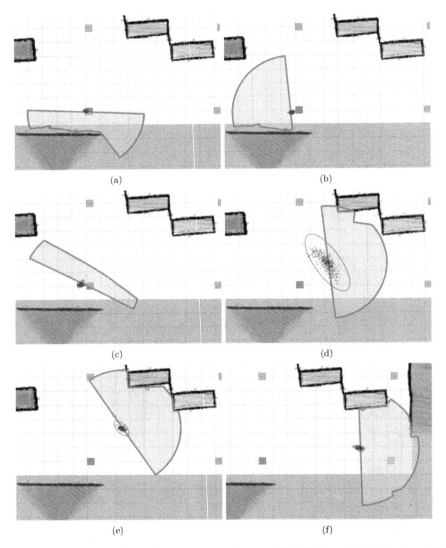

Figure 8.12: Screenshots during live execution, showing PPOMDP's typical behaviour. After reaching the lower-right goal in (a), PPOMDP turns right to maintain sight of the lower left features, as shown in (b). (c) shows the real-world complexity of the observation model: when accelerating, the robot sees the ground a few of metres ahead of it. (d) shows PPOMDP's strategy of heading to the north-east, trying to find a feature after leaving the lower-left goal. Since the laser scan is drawn from the mean of each belief, it is clear in (d) that a substantial difference has developed between the mean belief and the true pose. NONPROB has difficulty because it cannot anticipate this occurrence. When a feature is found, the robot will become well-localised as in (e). It can then proceed to the goal while remaining in sight of the features, as shown in (f).

tain, somewhere between the two southern goals. The reason for this is unclear, however one possibility is that the failures are a result of belief transitions which can occur in reality but which PPOMDP considers impossible. In particular, after wheeling right from the lower-left goal, PPOMDP hopes to see some of the features near the centre of the map. Usually it sees a feature, becomes well-localised, then turns and heads towards the lower-right goal. However sometimes it turns towards the lower-right goal just before observing a feature, presumably under the assumption that a feature will be observed as it turns. In the simulators this assumption is reasonable. In reality however, especially at ranges close to 8m, feature observations may not be made. If a feature observation is not made, the robot will have travelled a long distance and turned almost 360° without viewing a feature. Its heading uncertainty will be extremely large, such that it will be unable to reliably turn back. This analysis is supported by the fact that PPOMDP was more successful on the realistic simulator than in reality. A more accurate observation model would likely reduce the occurrence of this failure mode.

8.7 Summary

This chapter applied PPOMDP to scenarios of increasing complexity, culminating in online planning for navigation of real robots. Results showed that PPOMDP is capable of scaling to real problems, and that managing an entire distribution rather than simply the maximum-likelihood state can produce much more robust behaviour. PPOMDP's performance was shown numerically to be superior to that of a non-probabilistic planner. In addition, PPOMDP subjectively appears to be much more in control, as though it has a better understanding of the problem. To the author's knowledge, these results represent the most challenging robot navigation problem to which POMDP solution methods have successfully been applied to date.

Chapter 9

Conclusion

This book presented an approach to planning in partially observable continuous domains. The basic methodology was to consider the evolution of continuous probability distributions, or beliefs, parameterised by finite-length vectors of parameters. Fitted value iteration was the solution method adopted. Numerous improvements to this basic approach were presented, allowing the algorithm to scale to a real robot navigation problem.

This chapter proceeds as follows. Section 9.1 summarises the material presented in this book, Section 9.2 outlines possible avenues for future research, and Section 9.3 concludes.

9.1 Summary

The subject of decision making under uncertainty in continuous domains was introduced in Chapter 1. It was shown, in general terms, how the approach advocated in this book relates to the field. The main contributions of the book were summarised and an outline of the book was presented.

Chapter 2 reviewed the literature on POMDP solution methods in detail. It focussed particularly on approaches based on value iteration, and on the applicability of the various solution methods to continuous problems. It was shown that several methods from the literature can be seen as the application of fitted value iteration after the selection of a particular information space in which to represent histories of actions and observations.

Chapter 3 presented the approach advocated in this book, of applying fitted value iteration after selecting the space of Gaussian approximations. It motivated the choice of $\mathcal{I}_{gauss}$, and described the basic mechanics of what we call the PPOMDP algorithm. BlockWorld was introduced as a relatively simple continuous navigation problem on which to compare planning algorithms.

PPOMDP was compared against a state-of-the-art discrete POMDP solution method on this world, showing reasonable but not spectacular performance.

Results were improved in Chapter 4 by using Monte Carlo methods to construct a belief transition function. Several optimisations were presented such that transitions could be calculated efficiently, by re-using the calculations of predictions, expected observations, and likelihoods. The result was that plans required no more processing than for the simplified transition function presented in the previous chapter, but were of a much higher quality. Furthermore, it was shown that the complexity of each belief transition does not depend on the number of discrete states required to cover the state-space, implying that this approach will be capable of scaling to larger physical environments than approaches relying on an underlying discretisation.

Chapter 5 improved planning speed by showing how the calculation of the discrete belief transition function could be broken into two or more steps based on the assumption of conditional independence between observation components. The results showed that planning on Block-World could be performed in approximately one third of the time.

Until this point, it was required that the set of beliefs over which PPOMDP planned lay on a regular grid over belief-space. Chapter 6 relaxed this requirement, allowing the use of arbitrary belief sets. In order to do so, an efficient weighting function based on metric indexing schemes was introduced. The use of arbitrary belief sets was shown to result in a significant increase in scalability, speeding up planning and reducing memory requirements.

Chapter 7 integrated online, local forward search with offline, global value iteration. It reviewed the relevant literature on forward search, showed how forward search can be implemented efficiently for the particle-based PPOMDP algorithm, and presented results to demonstrate its effectiveness. It was shown how forward planning can be used to locally "fill in the gaps" of a coarse global plan.

In Chapter 8, the material from previous chapters was brought to bear on a real-world problem. PPOMDP was applied to several progressively more challenging environments, culminating in a live demonstration on real robots. To the author's knowledge, this problem represents the most challenging robot navigation application to which POMDP solution methods have been applied. Results showed that PPOMDP was significantly more reliable than a non-probabilistic planner. By considering the entire distribution rather than simply the maximum-likelihood position, PPOMDP maintained localisation quality and avoided potentially dangerous scenarios.

9.2   Avenues for Future Research

While the algorithm presented in this book produced good results on a difficult problem, this section discusses potential improvements to the algorithm and the possibility of its application to even more challenging problem domains.

9.2.1 Dynamic or Unknown Environments

The POMDP formulation presented in this book assumes that the environment is both static and perfectly known. This assumption is often invalid. In robot navigation terms, the simultaneous localisation and mapping (SLAM) problem [35] provides a counter-example, where the robot must discover the environment online. The simultaneous localisation, mapping, and moving object tracking (SLAMMOT) problem [121] adds extra complexity by relaxing the requirement that the environment be static.

A principled approach to planning in dynamic or unknown environments is to simply augment the state vector, as is done for estimation in the SLAM and SLAMMOT problems. Unfortunately, assuming a Gaussian approximation, the dimensionality of Gaussians becomes large and varies with the number of features. It seems unlikely that current POMDP solution methods will scale to considering the evolution of beliefs over map states and the states of moving objects, in addition to vehicle states. Furthermore, the likelihood of future observations in unknown environments is particularly challenging to model, although some work has been done on modelling observation likelihoods in unknown environments [104]. A simplified approach to planning in unknown or dynamic environments, which avoids augmenting the state vector, would be valuable.

9.2.2 Further Application of POMDP Solution Methods

In the author's opinion, a challenge for the practical application of POMDP solution methods is to find the right problem domain. The problem domain must be sufficiently simple to allow an accurate model, while still being sufficiently uncertain to benefit from a POMDP solution.

If the problem domain is too complicated, it can be difficult to model. As was discussed in Section 4.2, one approach to estimation in difficult-to-model environments is to choose to incorporate only those aspects of observations which are reliable. An example of this is feature-based localisation which ignores negative information. Section 4.2 explained why this is an appropriate strategy for estimation, but not for planning. A planner must be able to anticipate the likelihood of future observations. The fact that a feature was not observed on this time-step is definitely pertinent to the likelihood of it being observed on the next time-step. The modelling inaccuracies for the RealWorld environment, discussed in Chapter 8, suggest that RealWorld is approaching the threshold of an environment that is too complicated.

Conversely, if the problem domain is too simple, an accurate model can be used. While not necessarily the case, the application of powerful estimation techniques is likely to keep uncertainty small. For small enough uncertainty, a POMDP solution will be identical to a maximum-likelihood or heuristic solution. The counter-example to this scenario is a world which can be accurately modelled, but in which observations are either extremely uncertain or infrequent. This would represent the perfect scenario for the application of POMDP solution methods, since the likelihood of future observations could be accurately predicted, but high uncertainty would be problematic for heuristics.

9.2.3 Increased Belief Complexity

While a number of opportunities for optimisation were noted throughout this book, the most promising area for improvement is perhaps the extension from unimodal Gaussians to more complex probability distributions. Section 3.2 discussed the quality of a Gaussian approximation, and hence the importance of such an extension. It argued that Gaussians are a good model for most of the beliefs that occur during robot navigation problems. Multimodal distributions occur relatively infrequently (assuming that global localisation is an infrequent event), and present a less serious problem for an unprepared planner than for an unprepared localiser. However, non-Gaussian unimodal beliefs are more frequent. Section 8.6.1 discussed one such issue, involving an uncertain belief near a thin wall. Since the step change at the wall cannot be modelled by a single Gaussian, the planner believed that there was a non-zero probability of "tunnelling" through that wall.

This section suggests two approaches to handling non-Gaussian beliefs. The first is based on heuristics, while the second involves operating in a more expressive information space.

9.2.4 Heuristics for Dealing with Non-Gaussian Beliefs

While the Gaussian-based planner discussed in this book can only reason about future Gaussian beliefs, a relatively simple extension may allow it to reason about one non-Gaussian step. As discussed in Section 8.2, the online estimator used for belief tracking in Chapter 8 clusters particles before estimating Gaussians, producing a Mixture of Gaussians (MoG) representation. When multiple Gaussian components are reported, the approach taken has been simply to select the most likely component, and assume that this represents the true belief.

One can draw parallels between uncertainty about a discrete set of states, and uncertainty about a discrete set of Gaussian components. The approach of having blind faith in the most likely component is analogous to the MLS heuristic for dealing with POMDPs, discussed in Section 2.7.1, which has blind faith in the most likely state. Section 2.7.1 also described a set of more sophisticated heuristics, which may produce better results.

In particular, a version of the Q_{MDP} heuristic could be applied to MoG beliefs. Rather than acting according to the most likely Gaussian, each component of the MoG could vote for actions in proportion to the value of actions from that Gaussian belief, with a number of votes proportional to the component's weight. Q_{MDP} implicitly assumes that all state uncertainty will disappear after the next action and observation, and is optimal when that assumption is correct. When applied to MoGs, the heuristic would assume that the belief will collapse to a single Gaussian after the next action and observation.

Another option would be to generalise a dual-mode heuristic, such as the *Action entropy* heuristic described in Section 2.7.1. This would involve executing the PPOMDP plan as normal for a belief consisting of a single component, but taking actions to reduce the number of components when the uncertainty over components is beyond a threshold.

The heuristics described in Section 2.7.1 are unable to make long term plans which reason about how uncertainty will evolve, but can often deal with uncertainty over a short time-horizon. The Gaussian planner described in this book is able to make long term plans, but only about Gaussian uncertainty. The more non-Gaussian the true uncertainty, the more approximate these plans will be. However, the addition of heuristics for dealing with multi-modal beliefs will hopefully allow the planner to deal with non-Gaussian beliefs much more accurately over a short time horizon.

A potential complication for generalising Q_{MDP} is that the best actions from each Gaussian component must be considered. If forward planning is used, this is potentially expensive. The optimisations discussed in Section 7.6 may alleviate this expense. Another alternative is to reduce the amount of forward planning when beliefs are represented by multiple components.

9.2.5 Operating in a More Expressive I-Space

A more complete alternative to heuristics would be to operate in the space of mixtures of Gaussians. The ability to work with arbitrary belief sets would be critical, because the increase in the size of the parameter space would require an infeasibly large number of regular grid-points to cover it. As pointed out in the introduction to this chapter, the use of arbitrary beliefs means that scalability becomes limited by the size of the set of likely beliefs, rather than the size of the parameter-space in which beliefs are described. This is important, because the set of MoGs encountered during robot navigation is likely to be highly constrained.

The main impediment to planning with an arbitrary set of MoG beliefs is the choice of an appropriate distance metric. Comparing MoGs using the Parameter-Euclidean distance is inappropriate because the length of parameter vectors would be variable, and results would differ depending on the order in which individual Gaussian components are listed. A more appropriate choice may be a distance metric based on the underlying probability distribution described by the parameter vector. The analysis and results presented in Chapter 6 may provide a starting point.

Another complication of using mixtures of Gaussians is that the space of relevant beliefs is likely to be significantly larger. However, it may be possible to reduce the amount of computation by recognising that many of the calculations required to estimate transitions from Gaussian A and transitions from Gaussian B are the same as those required to estimate transitions from a mixture of Gaussians A and B. In other words, it may be possible to combine calculations from individual Gaussian components to avoid performing an entire set of calculations from each individual MoG belief.

Alternatively, it may be the case that a mixture of Gaussians is not the best representation. For certain problems at least, an entirely different parameterisation, such as the use of wavelet coefficients, may be more expressive and compact. In analogy to Roy's work on belief compression [94], it may be possible to learn a good continuous representation based on a set of sample beliefs. This would free the human designer from the responsibility of selecting a good representation for beliefs.

9.3 Conclusion

This book has contributed an algorithm for planning in partially-observable domains, which operates by planning in the space of continuous parameterised probability distributions. While the approach may not be appropriate for all domains, it was shown that good plans can be generated quickly when the structure of the domain is such that beliefs are usually well approximated by a particular parametric form. It was shown that robot navigation problems can involve sufficiently structured beliefs as to be amenable to this approach. A number of novel improvements to the basic algorithm were presented, to the point where the algorithm could solve challenging real-world problems and be implemented on real robots. This demonstration, involving modelling both position and heading, represents a significant improvement on the state of the art.

Appendix A

Derivation of w_μ for Repairing the Matusita Distance

Section 6.4.1 described a repaired Matusita distance metric, denoted D_{M_R}, such that

$$D_{M_R} = D_M + D_\mu \tag{A.1}$$

where D_M is the Matusita distance metric, and

$$D_\mu(\mathbf{v}_1, \mathbf{v}_2) = w_\mu \left\{ (\boldsymbol{\mu}_1 - \boldsymbol{\mu}_2)^T (\boldsymbol{\mu}_1 - \boldsymbol{\mu}_2) \right\}^{1/2} \tag{A.2}$$

This appendix provides a detailed derivation of w_μ.

Let $\max(D_M)$ denote the steady-state value which D_M approaches as the overlap between two distributions approaches zero. To ensure that D_μ only dominates for near neighbours, w_μ is chosen such that

$$D_\mu(\mathbf{v}_1, \mathbf{v}_2) > D_M(\mathbf{v}_1, \mathbf{v}_2) \quad \text{if and only if} \quad D_M(\mathbf{v}_1, \mathbf{v}_2) > s \max(D_M) \tag{A.3}$$

where s is a value close to 1, such as 0.9. In other words, the relative weighting is chosen such that D_M dominates until it approaches its steady-state value, at which point D_μ takes over.

The steady-state value of D_M is simple to evaluate from Equation 6.16. Since the Bhattacharyya distance is unbounded, the maximum value of D_M is approached as $D_B \to \infty$, and hence $\exp(-D_B) \to 0$. Therefore $\max(D_M) = \sqrt{2}$.

It is not possible to satisfy Equation A.3 for all $\mathbf{v}_1$ and $\mathbf{v}_2$, given the different shapes of the two metrics. Therefore the approach taken is to choose a representative covariance value and calculate an analytical $\tilde{w}_\mu$ such that Equation A.3 holds for two Gaussians with this covariance.

w_μ is then set using

$$w_\mu = M\tilde{w}_\mu \tag{A.4}$$

where M is a constant selected experimentally, by evaluating the efficiency of the resultant metric indexing structure versus the extent to which neighbours are re-ordered. The chosen representative covariance is the identity matrix $\mathbf{I}$.

Let $\mathbf{v}_1 = <\boldsymbol{\mu}_1, \Sigma_1>$ and $\mathbf{v}_2 = <\boldsymbol{\mu}_2, \Sigma_2>$, where $\Sigma_1 = \Sigma_2 = \Sigma$, and let

$$D_M(\mathbf{v}_1, \mathbf{v}_2) = s\max(D_M) \tag{A.5}$$

Therefore

$$s\max(D_M) = \left\{ 2\left[1 - \exp(-D_B(\mathbf{v}_1, \mathbf{v}_2))\right] \right\}^{1/2} \tag{A.6}$$

$$D_B(\mathbf{v}_1, \mathbf{v}_2) = -\log\left(1 - \frac{1}{2}\left[s\max(D_M)\right]^2\right) \tag{A.7}$$

From Equation 6.14, the Bhattacharyya distance between Gaussians of equal covariance is

$$D_B(p_1(\mathbf{x}), p_2(\mathbf{x})) = \frac{1}{8}(\boldsymbol{\mu}_1 - \boldsymbol{\mu}_2)^T \Sigma^{-1}(\boldsymbol{\mu}_1 + \boldsymbol{\mu}_2) \tag{A.8}$$

Substituting this into Equation A.7 gives

$$(\boldsymbol{\mu}_1 - \boldsymbol{\mu}_2)^T \Sigma^{-1}(\boldsymbol{\mu}_1 - \boldsymbol{\mu}_2) = -8\log\left(1 - \frac{1}{2}\left[s\max(D_M)\right]^2\right) \tag{A.9}$$

Taking square-roots of both sides,

$$\left\{ (\boldsymbol{\mu}_1 - \boldsymbol{\mu}_2)^T \Sigma^{-1}(\boldsymbol{\mu}_1 - \boldsymbol{\mu}_2) \right\}^{1/2} = \left\{ -8\log\left(1 - \frac{1}{2}\left[s\max(D_M)\right]^2\right) \right\}^{1/2} \tag{A.10}$$

Using a representative covariance value of $\Sigma = \mathbf{I}$,

$$\left\{ (\boldsymbol{\mu}_1 - \boldsymbol{\mu}_2)^T (\boldsymbol{\mu}_1 - \boldsymbol{\mu}_2) \right\}^{1/2} = \left\{ -8\log\left(1 - \frac{1}{2}\left[s\max(D_M)\right]^2\right) \right\}^{1/2} \tag{A.11}$$

To satisfy Equation A.3, it is required that

$$D_\mu(\mathbf{v}_1, \mathbf{v}_2) = D_M(\mathbf{v}_1, \mathbf{v}_2) \tag{A.12}$$

Combining Equations A.12, A.2, and A.5 gives

$$\tilde{w}_\mu \left\{ (\boldsymbol{\mu}_1 - \boldsymbol{\mu}_2)^T (\boldsymbol{\mu}_1 - \boldsymbol{\mu}_2) \right\}^{1/2} = s\max(D_M) \tag{A.13}$$

Substituting Equation A.11 and rearranging gives

$$\tilde{w}_\mu = s\max(D_M)\left\{-8\log\left(1 - \frac{1}{2}\left[s\max(D_M)\right]^2\right)\right\}^{-1/2} \tag{A.14}$$

Substituting $\max(D_M) = \sqrt{2}$ and simplifying produces

$$\tilde{w}_\mu = s\sqrt{2}\left\{-8\log(1 - s^2)\right\}^{-1/2} \tag{A.15}$$

Combining Equations A.15 and A.4 gives

$$w_\mu = Ms\sqrt{2}\left\{-8\log(1 - s^2)\right\}^{-1/2} \tag{A.16}$$

M was selected using a database of the same size and distribution as the database from Section 6.4. 5000 query points were randomly selected from the same distribution. For a set of metrics, the nearest-neighbour of each query point was calculated and the average number of required distance calculations was recorded. The lists of nearest neighbours were then compared, and the percentage of queries on which they agreed was calculated.

The number of comparisons and the rate of agreement represent competing objectives. A good compromise was found to be $M = 0.25$. The results for this choice of M are shown in Table 6.3.

Bibliography

[1] D. Aberdeen and J. Baxter. Scaling internal-state policy-gradient methods for POMDPs. In *Proc. Intl. Conf. on Machine Learning*, pages 3–10, 2002.

[2] C. Andrieu, N. de Freitas, A. Doucet, and M. Jordan. An introduction to MCMC for machine learning. *Machine Learning*, 50:5–43, 2003.

[3] K. Arras, J. Castellanos, M. Schilt, and R. Siegwart. Feature-based multi-hypothesis localization and tracking using geometric constraints. *Robotics and Autonomous Systems*, 44(1):41–53, 2003.

[4] S. Arulampalam, S. Maskell, N. Gordon, and T. Clapp. A tutorial on particle filters for on-line non-linear/non-gaussian Bayesian tracking. *IEEE Transactions on Signal Processing*, 50(2):174–188, 2002.

[5] H. Attias. Planning by probabilistic inference. In *International Conference on Artificial Intelligence and Statistics*, 2003.

[6] J. Bagnell and J. Schneider. Autonomous helicopter control using reinforcement learning policy search methods. In *Proc. IEEE Intl. Conf. on Robotics and Automation*, pages 1615–1620, 2001.

[7] T. Bailey and H. Durrant-Whyte. Simultaneous localisation and mapping (SLAM): Part II - state of the art. *Robotics and Automation Magazine*, 13(3):108–117, 2006.

[8] L. Baird and A. Moore. Gradient descent for general reinforcement learning. In *Advances in Neural Information Processing Systems*, volume 11, pages 968–974, 1999.

[9] S. Berchtold, D. Keim, and H. Kriegel. The X-tree: An index structure for high-dimensional data. In *Proc. 22nd Conf. on Very Large Databases*, pages 28–39, 1996.

[10] D. Bertsekas. *Dynamic Programming and Optimal Control*, volume 1. Athena Scientific, Belmont, Massachusetts, 2000.

[11] A. Bhattacharyya. On a measure of divergence between two statistical populations defined by their probability distributions. *Bull. Calcutta Math Soc.*, 35:99–109, 1943.

[12] J. Blythe. An overview of planning under uncertainty. *AI Magazine*, 20(2):37–54, 1999.

[13] C. Bohm, S. Berchtold, and D. Keim. Searching in high-dimensional spaces – index structures for improving the performance of multimedia databases. *ACM Computing Surveys*, 33(3):322–373, 2001.

[14] B. Bonet. An epsilon-optimal grid-based algorithm for Partially Observable Markov Decision Processes. In C. Sammut and A. Hoffmann, editors, *Proc. 19th International Conf. on Machine Learning*, pages 51–58, 2002.

[15] B. Bonet and H. Geffner. Planning with incomplete information as heuristic search in belief space. In *Proc. Intl. Conf. on AI Planning and Scheduling*, pages 52–61, 2000.

[16] C. Boutilier, T. Dean, and S. Hanks. Decision-theoretic planning: Structural assumptions and computational leverage. *Journal of Artificial Intelligence Research*, 11:1–94, 1999.

[17] R. Brafman. A heuristic variable grid solution method for POMDPs. In *Proc. Fourteenth National Conference on Artificial Intelligence (AAAI'97)*, pages 727–733, 1997.

[18] D. Braziunas and C. Boutilier. Stochastic local search for POMDP controllers. In *Proc. Nineteenth National Conference on Artificial Intelligence (AAAI-04)*, pages 690–696, 2004.

[19] A. Brooks, A. Makarenko, S. Williams, and H. Durrant-Whyte. Parametric POMDPs for planning in continuous state spaces. *Robotics and Autonomous Systems*, 54(11):887–897, 2006.

[20] A. Brooks, T. Kaupp, A. Makarenko, S. Williams, and A. Oreback. Orca: A component model and repository. In D. Brugali, editor, *Software Engineering for Experimental Robotics*, volume 30 of *Springer Tracts in Advanced Robotics*, pages 231–251. Springer, 2007.

[21] W. Burgard, D. Fox, D. Hennig, and T. Schmidt. Estimating the absolute position of a mobile robot using position probability grids. In *Proc. 13th National Conference on Artificial Intelligence*, pages 896–901, 1996.

[22] M. Campbell, A. Hoane, and F.-H. Hsu. Deep blue. *Artificial Intelligence*, 134(1-2): 57–83, 2002.

[23] A. Cassandra. *Exact and Approximate Algorithms for Partially Observable Markov Decision Processes*. PhD thesis, Brown University, Department of Computer Science, 1998.

[24] A. Cassandra, L. Kaelbling, and J. Kurien. Acting under uncertainty: Discrete Bayesian models for mobile robot navigation. In *Proc. IEEE/RSJ Intl. Conf. on Intelligent Robots and Systems*, volume 2, pages 963 – 972, 1996.

[25] A. Cassandra, M. Littman, and N. Zhang. Incremental pruning: A simple, fast, exact method for partially observable Markov decision processes. In *Proc. Conf. on Uncertainty in Artificial Intelligence*, pages 54–61, 1997.

[26] E. Chavez, G. Navarro, R. Baeza-Yates, and J. Marroquin. Searching in metric spaces. *ACM Computing Surveys*, 33(3):273–321, 2001.

[27] H. Cheng. *Algorithms for Partially Observable Markov Decision Processes*. PhD thesis, University of British Columbia, 1988.

[28] P. Ciaccia, M.Patella, and P. Zezula. M-tree: An efficient access method for similarity search in metric spaces. In *Proc. Intl. Conf. on Very Large Data Bases*, pages 426 – 435, 1997.

[29] K. Clarkson. Nearest-neighbor searching and metric space dimensions. In G. Shakhnarovich, T. Darrell, and P. Indyk, editors, *Nearest-Neighbor Methods for Learning and Vision: Theory and Practice*, pages 15–59. MIT Press, 2006.

[30] T. Cover and A. Thomas. *Elements of Information Theory*. John Wiley & Sons, 1991.

[31] S. Davies. Multidimensional triangulation and interpolation for reinforcement learning. In *Advances in Neural Information Processing Systems*, volume 9, pages 1005–1011, 1997.

[32] S. Davies, A. Ng, and A. Moore. Applying online-search techniques to continuous-state reinforcement learning. In *AAAI-98*, pages 753–760, 1998.

[33] R. Duda, P. Hart, and D. Stork. *Pattern Classification*. John Wiley and Sons, 2001.

[34] H. Durrant-Whyte. An autonomous guided vehicle for cargo handling applications. *Intl. Journal of Robotics Research*, 15(5):407–440, 1996.

[35] H. Durrant-Whyte and T. Bailey. Simultaneous localisation and mapping (SLAM): Part I - the essential algorithms. *Robotics and Automation Magazine*, 13(2):99–110, 2006.

[36] J. Eagle. The optimal search for a moving target when the search path is constrained. *Operations Research*, 32(5):1107–1115, 1984.

[37] A. Elfes. *Occupancy Grids: A Probabilistic Framework for Robot Perception and Navigation*. PhD thesis, Department of Electrical and Computer Engineering, Carnegie Mellon University, Pittsburgh, PA., 1989.

[38] C. Faloutsos, W. Equitz, M. Flickner, W. Niblack, D. Petkovic, and R. Barber. Efficient and effective querying by image content. *Journal of Intelligent Information Systems*, 3 (3/4):231–262, 1994.

[39] R. Fikes and N. Nilsson. STRIPS: A new approach to the application of theorem proving to problem solving. *Artificial Intelligence*, 2(3-4):189–208, 1971.

[40] R. Fikes, P. Hart, and N. Nilsson. Learning and executing generalized robot plans. *Artificial Intelligence*, 3(4):251–288, 1972.

[41] A. Foka and P. Trahanias. Real-time hierarchical POMDPs for autonomous robot navigation. In *IJCAI Workshop Reasoning with Uncertainty in Robotics*, 2005.

[42] D. Fox. Adapting the sample size in particle filter through KLD-sampling. *Intl. Journal of Robotics Research*, 22(12):985–1003, 2003.

[43] D. Fox, W. Burgard, and S. Thrun. Active markov localization for mobile robots. *Robotics and Autonomous Systems*, 25:195–207, 1998.

[44] D. Fox, W. Burgard, F. Dellaert, and S. Thrun. *Sequential Monte Carlo Methods in Practice*, chapter Particle filters for mobile robot localization, pages 401–426. Springer-Verlag, New York, 2001.

[45] J. Friedman, J. Bentley, and R. Finkel. An algorithm for finding best matches in logarithmic expected time. *ACM Transactions on Mathematical Software*, 3(3):209–226, 1977.

[46] V. Gaede and O. Gunther. Multidimensional access methods. *ACM Computing Surveys*, 30(2):170–231, 1998.

[47] H. Geffner and B. Bonet. Solving large POMDPs by real time dynamic programming. In *Proc. Fall AAAI Symposium on POMDPs*, pages 61–68, 1998.

[48] B. Gerkey, R. Vaughan, and A. Howard. The Player/Stage project: Tools for multi-robot and distributed sensor systems. In *Proc. Intl. Conf. on Advanced Robotics*, pages 317–323, 2003.

[49] G. Gordon. Stable function approximation in dynamic programming. In *Proc. Intl. Conf. on Machine Learning*, pages 261–268, 1995.

[50] J. Guivant and E. Nebot. Solving computational and memory requirements of feature-based simultaneous localization and mapping algorithms. *IEEE Transactions on Robotics and Automation*, 19(4):749–755, 2003.

[51] A. Gutman. R-trees: A dynamic index structure for spatial searching. In *Proc. ACM SIGMOD Intl. Conf. on Management of Data*, pages 47–57, 1984.

[52] E. Hansen. Solving POMDPs by seraching in policy space. In *UAI 1998*, pages 211–219, 1998.

[53] M. Hauskrecht. Value-function approximations for partially observable Markov decision processes. *Journal of Artificial Intelligence Research*, 13:33–94, 2000.

[54] H. Neemuchwala A. O. Hero and P.L. Carson. Image matching using alpha-entropy measures and entropic graphs. *Signal Processing (Special Issue on Content-based Visual Information Retrieval)*, 85:277–296, 2005.

[55] G. Hjaltason and H. Samet. Index-driven similarity search in metric spaces. *ACM Transactions on Database Systems*, 28(4):517–580, 2003.

[56] J. Hoey and P. Poupart. Solving POMDPs with continuous or large discrete observation spaces. In *Proc. Intl. Joint Conf. on Artificial Intelligence*, pages 1332–1338, 2005.

[57] P. Jensfelt. *Approaches to Mobile Robot Localization in Indoor Environments*. PhD thesis, Royal Institute of Technology, Stockholm, 2001.

[58] P. Jensfelt and S. Kristensen. Active global localisation for a mobile robot using multiple hypothesis tracking. *IEEE Transations on Robotics and Automation*, 17:748–760, 2000.

[59] L. Kaelbling, M. Littman, and A. Moore. Reinforcement learning: A survey. *Journal of Artificial Intelligence Research*, 4:237–285, 1996.

[60] L. Kaelbling, M. Littman, and A. Cassandra. Planning and acting in partially observable stochastic domains. *Artificial Intelligence*, 101:99–134, 1998.

[61] S. Kakade. A natural policy gradient. In *Advances in Neural Information Processing Systems*, volume 14, pages 1531–1538, 2002.

[62] M. Kearns, Y. Mansour, and A. Ng. A sparse sampling algorithm for near optimal planning in large Markov decision processes. In *Proc. Intl. Joint Conf. on Artificial Intelligence*, pages 1324–1331, 1999.

[63] J.-C. Latombe. *Robot Motion Planning*. Kluwer Academic Publishers, 1991.

[64] S. LaValle. *Planning Algortihms*. Cambridge University Press, 2006.

[65] S. LaValle and S. Hutchinson. An objective-based framework for motion planning under sensing and control uncertainties. *Intl. Journal of Robotics Research*, 17(1):19–42, 1998.

[66] X. Li, W. Cheung, and J. Liu. Towards solving large-scale POMDP problems via spatio-temporal belief state clustering. In *IJCAI Workshop Reasoning with Uncertainty in Robotics*, 2005.

[67] M. Littman, A. Cassandra, and L. Kaelbling. Learning policies for partially observable environments: Scaling up. In *Proc. Intl. Conf. on Machine Learning*, pages 362–370, 1995.

[68] W. Lovejoy. Computationally feasible bounds for partially observed Markov decision processes. *Operations Research*, 39(1):162–175, 1991.

[69] D. Mackay. *Information Theory, Inference, and Learning Algorithms*. Cambridge University Press, 2003.

[70] K. Matusita. Decision rules, based on the distance, for problems of fit, two samples, and estimation. *Annals of Mathematical Statistics*, 26(4):631–640, 1955.

[71] D. McAllester and S. Singh. Approximate planning for factored POMDPs using belief state simplification. In *Proc. Conf. on Uncertainty in Artificial Intelligence*, pages 409–416, 1999.

[72] N. Meuleau, L. Peshkin, K. Kim, and L. Kaelbling. Learning finite-state controllers for partially observable environments. In *Proc. Conf. on Uncertainty in Artificial Intelligence*, pages 427–436, 1999.

[73] G. Monahan. A survey of partially observable markov decision processes: Theory, models and algorithms. *Management Science*, 28(1):1–16, 1982.

[74] D. Moore. *Simplicial Mesh Generation with Applications*. PhD thesis, Report no. 92-1322, Cornell University, 1992.

[75] R. Munos and A. Moore. Variable resolution discretization in optimal control. *Machine Learning*, 49, Numbers 2/3:291–323, November/December 2002.

[76] A. Ng and M. Jordan. PEGASUS: A policy search method for large MDPs and POMDPs. In *Proc. Conf. on Uncertainty in Artificial Intelligence*, pages 406–415, 2000.

[77] A. Ng, R. Parr, and D. Koller. Policy search via density estimation. In *Advances in Neural Information Processing Systems*, volume 12, pages 1022–1028, 1999.

[78] N. Nilsson. *Principles of Artificial Intelligence*. Tioga, Palo Alto, 1980.

[79] I. Nourbakhsh, R. Powers, and S. Birchfield. Dervish: An office-navigating robot. *AI Magazine*, 16(2):53–60, 1995.

[80] S. Omohundro. Five balltree construction algorithms. Technical report, International Computer Science Institute, TR-89-063, 1989.

[81] S. Paquet, L. Tobin, and B. Chaib-draa. An online POMDP algorithm for complex multiagent environments. In *Proc. Intl. Joint Conf. on Autonomous Agents & Multi-Agent Systems (AAMAS)*, pages 970–977, 2005.

[82] R. Parr and S. Russell. Approximating optimal policies for partially observable stochastic domains. In *Proc. Intl. Joint Conf. on Artificial Intelligence*, pages 1088–1095, 1995.

[83] J. Pearl. *Probabilistic Reasoning in Intelligent Systems: Networks of Plausible Inference.* Morgan Kaufmann, San Mateo, California, 1988.

[84] J. Pineau. *Tractable Planning Under Uncertainty: Exploiting Structure.* PhD thesis, Robotics Institute, Carnegie Mellon University, 2004.

[85] J. Pineau, G. Gordon, and S. Thrun. Applying metric-trees to belief-point POMDPs. In *Advances in Neural Information Processing Systems*, 2003.

[86] J. Pineau, G. Gordon, and S. Thrun. Point-based value iteration: An anytime algorithm for POMDPs. In *Proc. Intl. Joint Conf. on Artificial Intelligence*, pages 1025–1032, 2003.

[87] J. Pineau, G. Gordon, and S. Thrun. Policy-contingent abstraction for robust robot control. In *Proc. Conf. on Uncertainty in Artificial Intelligence*, pages 477–484, 2003.

[88] K. Poon. A fast heuristic algorithm for decision-theoretic planning. Master's thesis, The Hong Kong University of Science and Technology, 2001.

[89] J. Porta, N. Vlassis, M. Spaan, and P. Poupart. Point-based value iteration for continuous POMDPs. *Journal of Machine Learning Research*, pages 2329–2367, 2006.

[90] P. Poupart and C. Boutilier. Value-directed compression of POMDPs. In *Advances in Neural Information Processing Systems*, pages 1547–1554, 2002.

[91] P. Poupart and C. Boutilier. Bounded finite state controllers. In *Advances in Neural Information Processing Systems*, 2003.

[92] N. Roy. *Finding Approximate POMDP Solutions Through Belief Compression.* PhD thesis, Robotics Institute, Carnegie Mellon University, 2003.

[93] N. Roy and S. Thrun. Coastal navigation – mobile robot navigation with uncertainty in dynamic environments. In *Proc. IEEE Intl. Conf. on Robotics and Automation*, pages 35–40, 1999.

[94] N. Roy, G. Gordon, and S. Thrun. Finding approximate POMDP solutions through belief compression. *Journal of Artificial Intelligence Research*, 23:1–40, 2005.

[95] S. Russell and P. Norvig. *Artificial Intelligence: A Modern Approach.* Prentice Hall, 1995.

[96] H. Samet. The quadtree and related hierarchical data structures. *ACM Computing Surveys*, 16(2):187–260, 1984.

[97] B.W. Silverman. *Density Estimation for Statistics and Data Analysis.* Chapman and Hall, 1986.

[98] R. Simmons and S. Koenig. Probabilistic robot navigation in partially observable environments. In *Proc. Intl. Joint Conf. on Artificial Intelligence*, pages 1080 – 1087, July 1995.

[99] S. Singh, T. Jaakkola, and M. Jordan. Reinforcement learning with soft state aggregation. In *Advances in Neural Information Processing Systems*, pages 359–368, 1995.

[100] R. Smallwood and E. Sondik. The optimal control of partially observable markov processes over a finite horizon. *Operations Research*, 21(5):1071–1088, 1973.

[101] T. Smith and R. Simmons. Heuristic search value iteration for POMDPs. In *Proc. Conf. on Uncertainty in Artificial Intelligence*, pages 520–527, 2004.

[102] E. Sondik. *The Optimal Control of Partially Observable Markov Processes*. PhD thesis, Stanford University, 1971.

[103] M. Spaan and N. Vlassis. Perseus: Randomized point-based value iteration for POMDPs. *Journal of Artificial Intelligence research*, 24:195–220, 2005.

[104] B. Stewart, J. Ko, D. Fox, and K. Konolige. The revisiting problem in mobile robot map building: A hierarchical Bayesian approach. In *Proc. Conf. on Uncertainty in Artificial Intelligence*, pages 551–558, 2003.

[105] R. Sutton and A. Barto. *Reinforcement Learning: An Introduction*. MIT Press, Cambridge MA, 1998.

[106] G. J. Tesauro. Temporal difference learning and TD-gammon. *Communications of the ACM*, 38:58–68, 1995.

[107] G. Theocharous and S. Mahadevan. Approximate planning with hierarchical partially observable Markov decision process models for robot navigation. In *Proc. IEEE Intl. Conf. on Robotics and Automation*, pages 1347–1352, 2002.

[108] S. Thrun. Monte carlo POMDPs. In *Advances in Neural Information Processing Systems*, pages 1064–1070, 2000.

[109] S. Thrun. Particle filters in robotics. In *Proc. Conf. on Uncertainty in Artificial Intelligence*, pages 511–51, 2002.

[110] S. Thrun, D. Fox, W. Burgard, and F. Dellaert. Robust monte carlo localization for mobile robots. *Artificial Intelligence*, 128(1-2):99–141, 2001.

[111] S. Thrun, W. Burgard, and D. Fox. *Probabilistic Robotics*. MIT Press, Cambridge, MA, 2005.

[112] B. Tovar, A. Yershova, J. O'Kane, and S. LaValle. Information spaces for mobile robots. In *International Workshop on Robot Motion and Control (RoMoCo)*, 2005.

[113] J. Tsitsiklis. Decentralized detection. *Advances in Signal Processing*, 2:297–344, 1993.

[114] J. Tsitsiklis and B. Van Roy. Feature-based methods for large-scale dynamic programming. *Machine Learning*, 22:59–94, 1996.

[115] E. Tuttle and Z. Ghahramani. Propagating uncertainty in POMDP value iteration with Gaussian processes. Technical report, Gatsby Computational Neuroscience Unit, University College London, 2004.

[116] J. Uhlmann. Satisfying general proximity/similarity queries with metric trees. *Information Processing Letters*, 40:175–179, 1991.

[117] I. Ulrich and J. Borenstein. VFH+: Reliable obstacle avoidance for fast mobile robots. In *Proc. IEEE Intl. Conf. on Robotics and Automation*, pages 1572–1577, 1998.

[118] R. Upton. *Dynamic Stochastic Control - A New Approach to Game Tree Searching*. PhD thesis, Warwick University, 1999.

[119] D. Verma and R. Rao. Planning and acting in uncertain environments using probabilistic inference. In *Proc. IEEE/RSJ Intl. Conf. on Intelligent Robots and Systems*, 2006.

[120] E. Vidal. New formulation and improvements of the nearest neighbours in approximating and eliminating search algorithm (AESA). *Pattern Recognition Letters*, 15(1):1–7, 1994.

[121] C.-C. Wang. *Simultaneous Localization, Mapping and Moving Object Tracking*. PhD thesis, Robotics Institute, Carnegie Mellon University, Pittsburgh, PA, 2004.

[122] R. Washington. Incremental Markov-model planning. In *Proceedings of TAI-96, Eighth IEEE International Conference on Tools With Artificial Intelligence*, pages 41–47, 1996.

[123] R. Washington. BI_POMDP: Bounded, incremental partially-observable Markov-model planning. *Lecture Notes in Computer Science*, 1348:440–451, 1997.

[124] R. Williams. Simple statistical gradient-following algorithms for connectionist reinforcement learning. In *Machine Learning*, volume 8, pages 229–256, 1992.

[125] P. Yianilos. Data structures and algorithms for nearest neighbor search in general metric spaces. In *Proceedings of the Fifth Annual ACM-SIAM Symposium on Discrete Algorithms (SODA)*, pages 311–321, 1993.

[126] R. Zhou and E. Hansen. An improved grid-based approximation algorithm for POMDPs. In *Proc. Intl. Joint Conf. on Artificial Intelligence*, pages 707–716, 2001.

[127] S. Zhou and R. Chellappa. From sample similarity to ensenble similarity: Probabilistic distance measures in reproducing kernel Hilbert space. *IEEE Transactions on Pattern Analysis and Machine Intelligence*, pages 917–929, 2006.